Dear Tom,
Great to meet you at
our NASA seminar. Keep
up the good work.

Leighton

FIRE, ICE AND PARADISE

H. Leighton Steward

authorHOUSE®

AuthorHouse™
1663 Liberty Drive
Bloomington, IN 47403
www.authorhouse.com
Phone: 1-800-839-8640

First published by AuthorHouse 5/20/2009

ISBN: 978-1-4389-8379-0 (sc)

For additional copies contact AuthorHouse Book Orders at: 1-888-280-7715

Printed in the United States of America
Bloomington, Indiana

This book is printed on acid-free paper.

Contents

FORWARD

Leighton Steward has a history of taking issues of seemingly considerable complexity and boiling them down to a few salient points that leave his readers more knowledgeable and enlightened. The task of reducing global climate change, which includes global warming, to a few key points for his intended audience, the layman, is particularly challenging, but I believe Leighton has accomplished it. In addition, he has included several summary tables and charts that will allow readers to follow the ever-evolving research results on past and present climate and will stimulate their curiosity and expand their knowledge of Earth's dynamic history. He chronicles that Earth was birthed in fire, almost frozen in ice, and is now in an under-appreciated paradise of a climate.

Growing up outside a small town in a rural setting, Leighton had daily exposure to the fields and the woods and the animals that lived therein. His affinity for this setting was probably what gave him an early awareness of Earth's more pristine environments. As the years sped by, Leighton left his home town, concrete replaced the grass, and skyscrapers supplanted the natural shade from the trees. But he obviously never lost his appreciation for his earlier environment.

He pursued a geology degree at Southern Methodist University and minored in geography, which introduced him to meteorology. Meteorology is a study of the weather and the factors that not only cause it but also cause it to change. His early interest in the climate has developed into a serious fascination with and study of Earth's earlier climates, its paleoclimates.

The paleoclimate interest came naturally in other ways. A Master of Science degree in geology provided him with the knowledge of other

Earth processes, both present and past. As you will see, many of these natural processes affect the climate. The climate had a profound effect on another of his early interests, the environment.

Following college, as reconnaissance leader of a photo-mapping squadron when he was an officer in the U.S. Air Force, Leighton spent a significant part of three years in the Amazon and Orinoco basins of South America traveling by foot, dugout canoe, burro, Jeep, or the occasional helicopter. What he witnessed was pristine; no clear-cutting of the forests - no cutting at all. Like most of us, he doesn't like what he sees happening there today.

As good fortune would have it, he later became head of the Louisiana Land and Exploration Company, an energy company that was also the largest private owner of coastal wetlands in the lower 48 states - 600,000 acres in all. Located in the Mississippi River delta, this fragile land is still under siege, fraught with the problems of land loss caused primarily by man's leveeing the huge river and starving the delta of new sediment. Leighton's first crack at authorship, with yeoman help from his wetlands manager, Bill Berry, resulted in a publication that explained exactly why the land loss was occurring; a true accounting that had not been picked up by any of the media. First published in 1988, *Louisiana' National Treasure* was a success and changed both the media's characterization and the public's opinion of why the wetlands were disappearing. If a lawmaker doesn't know what is causing a problem, how can he or she enact legislation to properly fix that problem? More than 110,000 copies of this publication now reside in high schools, universities, governmental agencies, and in the hands of other interested parties. I frequently hear that the secret of *Louisiana's National Treasures'* success is that he simplified it, made it easy to understand, and did not propagandize for any particular interest.

Because of his company's voluntary efforts in trying to save this huge wetlands system, as well as him recommending measures that would benefit all of the nation's wetlands, the Environmental Protection Agency presented his company and him with it's highest honor for environmental excellence.

Next came Leighton's work as the lead author of a book on how people should live and eat to lose weight and avoid debilitating illnesses, particularly diabetes. After sixty-two drafts and sales of 210,000 copies of a self-published edition, he and his coauthors turned it over to a national publisher and the book *Sugar Busters: Cut Sugar to Trim Fat* became an instant number one *New York Times* bestseller. More than four million copies of *Sugar Busters* and its related offspring have made it to the printing press. Interestingly, Leighton received the same comments as he did with *Louisiana's National Treasure*. The subject matter was presented in simple, easily understandable language. It was effective, and has been imitated by other authors and diet companies as well as the food manufacturing industry that now offers consumers the choice of whole grain, high fiber and no sugar added products.

Leighton's early career included a stint as a field leader for a project in the Gulf of Mexico that was drilling and coring sediments below the sea floor, taking water samples, and studying surface to sea floor current conditions on a truly ground breaking research project on the upper continental slope in 600 to 4,000 feet of water. This was from 1964 to 1966, when such a coring project had never been done in the Gulf. It taught Shell, his employer, about sediment and fossil types deposited far beyond the current coastline, those deposited nearby when sea level was low during the coldest parts of the glacial cycles and then the return to the present high sea level stand of this warm, interglacial period.

Years later, in studying the reasons for the loss of the lower Mississippi River wetlands system, the specter of climate change and rising sea level again became a key area of concern. Leighton was chairman of the *National Wetlands Coalition* for ten years where nearly every recommendation or decision was weighed against how to have a net gain in our country's wetlands. For five years he was chairman of the *Audubon Nature Institute.* He also got involved in other topics of national interest, such as finding a better way to obtain oil to fill the *Strategic Petroleum Reserve* and changing our government's nutritional guidelines. One-on-one presentations and recommendations were given to two EPA administrators, two secretaries of the Department of the Interior, two secretaries of the Department of Energy, and several senatorial and congressional committees. In addition, he was asked to be the spokesman for the independent sector of the energy industry on *presidential missions* to the then USSR and Turkey (for G. H. Bush) and Pakistan (for Clinton). Obviously, someone trusted his advice!

Leighton received several awards from the energy industry for his leadership in proactively and voluntarily spending (non-requisite) dollars and efforts in environmental areas. The most prestigious energy industry trade association, The American Petroleum Institute (API) presented him with their highest award, the Gold Medal for Distinguished Service, for being an exemplary leader in environmental practices.

With his educational background, lifelong interest in the environment, and an entire world wallowing in confusion about what is causing our Earth's climate to change, Leighton thought, why not "go for three?" Why can't global climate change be explained as clearly and simply as wetlands loss and the best diet for mankind? This is an extremely complex subject and not easily reducible to two or three one-liners as to what makes Earth's climate change. But who better than Leighton to do it, otherwise we may have to settle for an overly scientific version written by a well-meaning Ph.D. who may be an expert on the subject, but is so brilliant that most of the writing or presentation goes right over the non-scientists' heads.

I think Leighton is eminently qualified to write an objective, layman-level review of the complex interaction of factors that come together to cause global climate change.

Gene Shinn, Ph.D.
Courtesy Professor
Marine Science Center
University of South Florida

PREFACE

Fire, Ice and Paradise is meant to be an educational and interesting guide to global climate change. The intended audience includes the laymen and policymakers who need fact-based input to help them better understand the climate-change issues. Most media information is only someone's opinion or a sensationalized sound-bite and not based on scientific fact or even the best current science. The many summary tables and charts that are presented here are intended to provide everyone with a handy reference to look to as the climate-change issue continues to evolve.

The initial inspiration for writing this book came from a discussion with Dr. Crayton Yapp, professor of geological sciences, at my alma mater, Southern Methodist University. Dr. Yapp's research was indicating that the Earth's carbon dioxide (CO_2) levels were probably as much as 18 times higher 500 million years ago, during the Cambro-Ordovician Period, than the CO_2 levels of today. At the time, I was unaware that Earth's paleo (old) atmosphere had contained such high levels of this greenhouse gas.

I then began researching books, textbooks, and articles written by other experts in climate study and was further inspired by discussions with other professors at SMU and by remembering conversations or presentations on factors that cause climate change while I was a member of the Advisory Board at *Lamont-Doherty Earth Observatory* at Columbia University. Lamont had been the first institution to have great predictive success relative to future El Niño - La Niña Pacific Ocean circulation events, as well as acquiring the magnetic data and mapping the ocean floor to confirm the continental drift or sea floor spreading hypothesis.

Summarized here for the general public are the factors that cause or "drive" the climate as well as discussions of the frequent climate changes that have occurred and will continue to occur on Earth. These factors are numerous and their interactions can be very complex or, in some cases, fairly obvious and straightforward. I will attempt to limit most of my observations to the major generalizations that can help the reader get a better feel for what is going on and why. From time to time, examples will be cited of the complexities of climate studies that can drive even the best researchers and experts crazy as they attempt to understand the past, much less make reliable predictions about the future.

The information presented here should help you better understand and interpret what you hear every day and also enjoy or connect with it more. More knowledge will help you screen the often sensationalized media spins that are injected into otherwise straightforward stories or facts regarding the climate.

In the figures and charts, I will often use the abbreviations *bya* for billion of years ago, *mya* for millions of years ago, *kya* for thousands of years ago, and *ppm* for parts per million. GHG will be used for greenhouse gases and CO_2 for carbon dioxide, O_2 for oxygen, CH_4 for methane, and *E.T.* for extraterrestrial. Temperatures will be identified in either degrees Fahrenheit (F) or Celsius (C) and I have provided a comparative scale so you can translate the temperatures mentioned here, or those found in other publications, into the scale in which you are most comfortable (Figure 1). Distances are often given in miles or kilometers. One kilometer is approximately 0.6 mile or a mile is approximately 1.6 kilometers.

To fully appreciate the pace of Earth's historical evolution, I ask you to try and comprehend very, very long periods of time and very, very slow Earth dynamics, such as continental movements that are measured in one to three inches a year (Figure 2). For instance, one inch a year is a foot in 12 years, a mile in 63,360 years, and 100 miles in 6.3 million years. The current width of the Atlantic Ocean has been created in about 175 million years, as the Americas split and then drifted away from Europe and Africa.

Fig. 1   <u>CELSIUS–FAHRENHEIT SCALES</u>

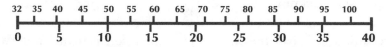

Degrees C (below) & Equivalent Degrees F (above)

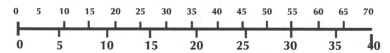

Degrees C (below) & Effect of Change in Degrees (above and vice-versa)

TO CALCULATE EQUIVALENT TEMPERATURES:
$$\text{DEGREES F} = \tfrac{9}{5}(Tc) + 32$$
$$\text{DEGREES C} = \tfrac{5}{9}(Tf - 32)$$

Modified from Farmers Almanac, *2007*

Fig. 2   <u>TO UNDERSTAND THIS SUBJECT BEST</u>

THINK IN VERY LONG PERIODS OF TIME

100 years ago — First airplanes and cars; pneumonia meant death

1,000 years ago — Medieval times, warmer than today

10,000 years ago — Earth warmed to current level from ice age

100,000 years ago — The last time Earth had warmed

1,000,000 years ago — Earth has had 10 ice ages since then

10,000,000 years ago — No permanant ice in Northern Hemisphere

100,000,000 years ago — Earth was 20° F warmer, CO_2 eight times higher

1,000,000,000 years ago — Was 1,000 million years ago!

4,600,000,000 years ago — Earth was formed

THINK SLOW

Earth's continents are drifting at one to three inches per year. At one inch, that is one foot in 12 years, one mile in 63,360 years, and 100 miles in 6.3 million years.

The shape of Earth's orbit around the sun changes from nearly circular to most elliptical and back every 100,000 years.

H. Leighton Steward, 2007

Relative to the amounts of the various gases that make up our atmosphere, I was advised by a former member of the United States Senate, when acknowledging that very few of our lawmakers have any significant background in science, that they are going to vote on technical issues in the way they or their constituents understand the issues. He continued that they certainly didn't know of what significance was a part per million or part per billion. Let's start with one part per hundred which is the same ratio of one penny to one dollar (100 pennies). One penny per 1,000 pennies (10 dollars) is still comprehensible. One part per million is one thousand thousand and one part per billion is one thousand million or a very small speck in the atmosphere. Whether that tiny speck is of significance depends on the impact it can impart. Many substances that are deadly at parts per hundred are totally benign at parts per million or billions. Arsenic, a naturally occurring "poison" in Earth's environment, is a good example. At a few parts per million or billion of arsenic, it has no poisonous effect on human life.

Without the training to know how many parts per whatever are significant, the lawmaker is still at the mercy of someone who can tell him or her what level is significant. If I wanted to stop the development of a facility that might release one part per billion of arsenic into a stream, I would just feed the media the arsenic word and scare the heck out of the voters and lawmakers and then sit back and watch the experts try and dig themselves out of the hole that had been created by my simply floating the arsenic word.

Where does that leave you, the non-scientist reader? Unfortunately it leaves you at the mercy of the individual feeding you the information. My best advice, when you hear of a coming catastrophe, is to get a second, non-biased opinion and ask them for some peer reviewed evidence to back up their position. This is particularly important when you hear only a one-liner summary of any impending catastrophe.

I sincerely hope you receive lasting value from reading this book, whether or not you agree with the conclusions I put forth. The opinions contained in this book are my own and should not be misconstrued as those of anyone I have identified in various ways in my text and acknowledgments.

So sit back, adjust the light just right, and allow me to share some brief remarks and a liberal use of summary charts and illustrations to see if we can reduce the key points about Earth's past and present climate to as few as ten. Review them while considering the scientific evidence or expert opinions contained in the text or illustrations.

INTRODUCTION

This book is a biography of the Earth's climate history summarized from research papers and works of many scientists and knowledgeable authors, plus information obtained from personal conversations with several of them. It also includes my own observations and thoughts. The audience I am targeting is _you_, whether you are a layperson or a lay scientist. If I cannot get you interested in global climate research, I will have to settle for making you the best informed person on the block regarding Earth's past and present climate. In your lifetime, this topic is not going to go away. If anything, it will become more and more a part of everyday conversation.

How the world addresses global climate change will be one of the most important topics of our lives. The many factors that determine climate change are driven by the various laws of science. The policies that are put into man's laws are determined by legions of non-scientists who are naïve regarding what is or is not causing climate change. Included in this group is a new president who will be extremely important in establishing the future policies and laws. Thus the need to get some "soft" science circulated to these key people as well as the general public that elects them. Otherwise, we may see a misdirection of our resources that will be spent dealing with only one narrow issue concerning global climate change.

I will discuss the major changes in Earth's climate, atmosphere, plant and animal life, and overall environment. The review begins a few billion years ago and continues right up to today. Relying on what is being reported in reputable scientific journals, university textbooks, and personal communications with well-known climate researchers, I also will tell you about the continuous small changes of more recent times to alert you to the fact that global climate change (GCC) is constant and that global climate stability is really a rare condition. Whether during the relatively warm intervals such as what we are experiencing today, or during the extended glacial cycles, climate has jumped up and down frequently, with many of the large changes happening over the course of only a few years or decades.

These constant climate changes, what we scientists call "chatter" or frequent changes which in many cases altered Earth's average temperature only a few degrees Fahrenheit, were sufficient to alter the course of many past civilizations. Examples will be given to substantiate that hypothesis. As we go farther back in time, the curves indicating greenhouse gas content and temperature are averaged (or "smoothed") because of the current sparsity of data and/or analyses. As additional data pours in, you will see that more climate variation or chatter will show up and wrinkle these smoothed curves.

Near the end of this book, I will recommend other writings that will be of great interest to those who want to know more about climate history and more about anthropology (the study of the history of humans) and how climate affected some of our predecessors.

This book contains a lot of science, but is written in layman's language as much as is possible. While I have attempted to summarize the better-known facts, if I should mention some less solid facts or speculate a little bit, I will identify my comments as such and use a "probably" or "possibly" to alert you to an as-yet-unproven point. When a proven answer does not exist, you often will get my informed opinion since I think I owe you that.

Most authors who write on subjects involving technical studies, proper nutritional guidelines, or, particularly, in voicing a specific political view, claim to have the exact answer to the subject matter at hand. This is not such a book! While I hope this is the best current summary, if not simplification, of this extremely complex subject, all the answers about why Earth's climate was like it was for the past 4.6 billion years do not yet exist.

Once you have digested the material presented herein, I hope you will be able to speculate on your own about how and why the climate changes or has changed and join me in looking forward to the new knowledge that is continuously becoming available in scientific literature. Trips to the post office that I used to dread for all the bills that were lying in wait now create anticipation for *Science, Nature, Geotimes*, the science sections of the newspaper; any articles that might contain a brand new discovery that can be plugged into the summary charts and tables contained in this book.

If you see a book, a film, or an article that claims to contain all of the answers to past and future climate change, be very skeptical, because no such thing is possible in our current state of knowledge. In fact, one of the goals of this book is to show that *there is no one answer*. Sometimes we are so hungry for answers that we accept certain conclusions too easily. So, again, be particularly skeptical when climate changes, such as global warming, are declared to be caused by a single factor, such as the currently popular belief that global warming is caused only by the increased CO_2 content in the atmosphere. CO_2 is only one factor in a complex system containing many active climate drivers.

I want you to get to know Earth better, its relationship to the sun and the moon, how and where its continents have moved, and particularly its climate history. Will our current paradise of a climate be preserved or become another paradise lost? I hope some of you will want to support, with your vote or influence (or even dollars), more basic research so we will be better prepared to try to extend the life of this current paradise and not see it disappear into the grips of another ice age like those that ended the earlier, comfortably warm, interglacial climates within the last one million years of Earth's history.

You will learn about volcanic lava flows, volcanic explosions, extraterrestrial impacts, drought and dust storms, continental flooding, and extreme ice cover (glaciations). All of these events, which have been portrayed as examples of abnormal weather or catastrophes, in fact have always occurred and quite frequently throughout time. The reason they seem more frequent and catastrophic now is because of the current 24/7 global media coverage, which was non-existent until very recently. Examination of official records and scientific indicators from decades past shows that the frequency of such events is really no greater today. It just seems that way.

Does this imply that we have had no effect on Earth's recent climate? No. If we put anything into the atmosphere or the oceans, or alter the presence or amount of forests, grasslands, marshlands, or anything that changes the reflectivity or heat absorption of the surface of the planet, we will change the overall environment in some minor and usually insignificant way. But if we add or alter something in a larger, more far-reaching way, we may instigate changes that do have a more measurable effect. For an illustration of our growing leverage to cause an effect, see Figure 3.

Will alterations caused by us, if we should be so lucky as to develop that power, be good or bad? It depends on what it does to you or to others that inhabit the planet. This impact would include not only humans but other fauna (animal life) and flora (plant life) as well. Our ultimate goal should be to not make alterations or changes that, while temporarily advantageous, eventually could lead to very unintended consequences such as causing Earth to get colder.

What is the best thing we can do to understand changes we might make that would result in a positive change or stabilization and, also, what might produce a negative result? We need to understand better what has occurred and where, when, and why, and then make an informed judgment about what to try to do or not do. How can we accumulate this knowledge? The sediments, rocks, and fossils have been waiting to tell their story of Earth's history for eons. This history includes information on Earth's climate, its inhabitants, the land and ocean's relative positions on Earth, and even evidence of "visitors" (meteors or comets) from outer space. Yes, there has been research, but scientists only recently have developed the technology and tools to decipher Earth's "code." We must go to Earth's "library" (see Chapter Two) and find out what all this new information means. With more studies of what is in Earth's "library", the data will expand and the resulting interpretations will become more precise.

If you are currently a student or teacher, I hope this will make you want to study some of the basic or peripheral sciences that apply to our climate or its history. Short of that, since the weather is the number one most-talked-about subject in conversations these days, you instantly will become one heck of a conversationalist!

If you enjoy solving puzzles, you should like this subject matter. The major causes and effects of climate change are defined and discussed in terms that will help most of you understand their meaning. However, the various causes and effects, which may be obvious and simple when analyzed separately, become much more complex when several of these drivers become active or reactive at the same time. Chapter Three on Climate Drivers contains a list of all the different pieces and how many of them interact. This will help you understand the reasons for the "heated" debates on what is currently causing global climate change.

Fortunately, for those of us who like to solve problems, a scientific consensus on many of the answers has not yet been established. Although you may not be a true scientist yourself, by becoming knowledgeable about the individual pieces, you may just gain an interesting insight regarding potential answers. As an informed citizen, you can share your questions with the experts and legislators who may not have looked at the climate changes from your particular perspective and your input might have implications for future policy making. "More heads are often better than one."

Fig. 3

400 B.C. TO 2,000 A.D.
GLOBAL POPULATION GROWTH

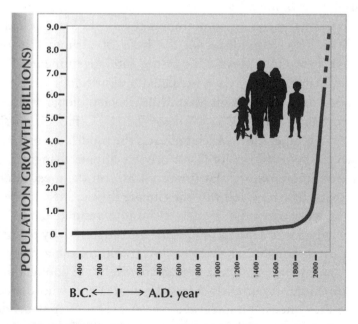

Nothing we do about our environment will make any difference in the long run if we don't address population. The U.S. Government says there will be nine billion people on Earth by 2050!

Gerhard, 2007

Also, to demonstrate that there really are many sources of data that allow these conclusions to be drawn from the past, I will summarize the data sources or paleoclimate indicators for you. These indicators will range in quality from excellent to only fair or directional and I have labeled the limitations of each in my summary chart.

I have not included a separate write-up on the intricacies of climate models, a complex subject in itself. Models can be helpful, even important. But while the science of modeling can produce some precise, reliable results in some fields, such as predictions of stress capacities for buildings or dams, models have been less precise in the area of paleoclimatology, and for good reason. Our understanding of the historic effects of individual climate drivers is only in its early stages, not to mention our understanding of the magnitude of the effects of many of these drivers and how they interact with other climate drivers. Models can be helpful, but my opinion of them is summarized in a general observation; nearly all climate models of the past three decades have been wrong in some significant way so, as of this writing, I can't comfortably recommend any current model as being highly reliable. Many reputable scientists from around the world are shocked to see key policies being considered on the basis of obviously inadequate models. For example, the models of the Intergovernmental Panel on Climate Change (IPCC) have predicted a continuous rise in Earth's temperature as CO_2 levels increase while, in fact, global temperatures have been falling since 2001. None of the IPCC models predicted a downturn in temperature, which indicates the models are programmed to show CO_2 being the major factor driving climate change.

We need more research (at the very least, on the most recent 500 million years) into how and why the climate drivers affected the earlier climates so we can load the models with more accurate data. When a factor is entered into a model to measure Earth's sensitivity to the factor, that can make some rational sense. But if a lot of the accompanying input is incomplete or imprecise, accurate results are doomed from the start. Also, there are currently at least 18 such factors whose effects are, at this point, somewhat imprecise, so the answer is often dependent on which factor or factors the modeler chooses to give the most weight. The proof of the pudding lies in the lack of success in predicting the past climates; imprecise data were entered, and/or key factors were left out.

Some of you are where I was only a few years ago when I knew very little about current climate change, much less about historical climate change and the factors that controlled it. Immerse yourself into this summary of climate changing factors or drivers and see if it will also "turn you on". As you read of the ancient times, you will wonder how in the devil we know about climates before people and their writings came into existence. Science and technology have presented us with just such an opportunity. And as I have previously said, they have allowed us in only the past few decades to unlock Earth's library so we can read from the rocks, sediments, trees, snow, ice, and pollen, etc. about these paleoclimates. Just a couple of decades back, global climate change was not even on the curriculum of most universities.

While global warming is getting the most time and coverage today, it is only one part of the much broader issue of global climate change which is the major focus of this book.

Chapter One

PARADISE IN PERSPECTIVE

Is today's moderately warm, calm, and relatively moist climate typical of Earth's historical climate? Not unless you define recent as only the last 10,000 years of the 4.6 billion years of Earth's existence. We are living in an occasional and brief flicker of comfortable, nourishing, and relatively stable climate. In researching the extremes of Earth's climate history, I realized how fortunate we are to be living at this particular time.

In writing about the history of Earth's climate and the many environments that have existed over the past 4.6 billion years, the title *Fire, Ice and Paradise* seemed to encompass the entire spectrum of our past. For the first few hundred million years, Earth was awash in the fire of molten material that covered its surface caused by the accumulation of the debris from space that formed our planet, plus the ensuing generation of tremendous heat from the constant impacts as these smaller bodies rained down at tens of thousands of miles per hour.

As Earth cooled, the surface became hospitable enough to harbor a very primitive life, a type of bacteria that can grow in very hot water. Nearly four billion years later, evidence on all of the continents indicates that Earth had glaciers and probably sea ice over most of its surface. Thus, the period, from about 850 million to 650 million years ago frequently has been referred to as the "Snowball Earth" time. Whether the planet was ever completely covered with ice has not been unequivocally proven but this was certainly a time of a vast amount of ice. There is evidence that a significant glaciation occurred as early as around 2.2 billion years ago. So Earth had, by 600 million years ago, certainly experienced fire and ice.

Everybody has his or her own idea of "paradise", but I think you will see that, certainly for humankind, this current climate and overall environment are as good as, if not better than, at any other time in Earth's history. No one would want to lose such an environment so let us try to understand the factors that have gotten our planet to this particular state and understand what we might be able to do or not do to try to preserve this paradise.

There were billions of years when humans could not have survived at all, either because of the climate or the critters. There were other periods when, had our species been present, people might have been able to survive, but it would have been in extreme discomfort. Today, the climate is such that people can survive to the ends of the Earth and comfortably thrive on most of the planet's landmasses. There are Eskimo (Inuit) people in the Artic, people living in the latitudes of four seasons, those existing in perpetually warm zones of the low latitudes, and even desert dwellers.

Since modern "man" has inhabited Earth, the global climate has been colder for most of that time. Many of our ancestors of more than 10,000 years ago lived in harsh glacial conditions for 90 percent of the time. I will review other intervals of time so you can see how our paradise compares. Figure 4, page 4, illustrates the time in Earth's history when the temperature, CO_2 and oxygen levels were generally higher or lower than the levels of today. It also indicates the glacial periods, times of vast desert deposits, sea levels, approximate locations of the continents and times of large extraterrestrial impacts. The text in Chapter Five will mention some of the dominant life forms.

Many of us would consider a Hawaiian type of climate as the very definition of paradise. It certainly is, but as you compare climates from most of Earth's multi-billion-year history, I think you will agree that nearly all of Earth's surface today, with the exception of the two poles and a few deserts, basks in a relative paradise. Today's global climate provides most of us with all the things we need to nurture ourselves, to flourish, and to prosper. Sure, we can have freezing-cold days and very windy days, but the annual averages are really quite nice. A large percentage of the surface waters of the planet's oceans are at a temperature that allows us to enjoy them in comfort and relative safety. We can even submerge ourselves in the spectacular real life aquariums using scuba devices.

Starting about 12,000 years ago and going back in Earth's history for the previous 85,000 to 90,000 years, this comfort zone would have been much narrower and closer to the equator, probably south from mid-Mexico latitudes. If our ancestors had had sailing ships, the higher average wind velocities of those cold times, with the accompanying higher waves, might have been fun for the hardiest sailors but the amateurs would have found it tough going and probably stayed on land.

Going back farther in time, for nearly a million years, these 85,000 to 90,000 years of cold, windy weather persisted for all but a handful of briefer 10,000 to 15,000 year interglacial intervals of relative warmth and calm. On land, the much drier climate accompanying these cold intervals would have caused all but a small percentage of a growing population to be underfed and undernourished.

Some say warmer climates bring on more diseases. Possibly, but tell that to the historians who have tracked the "Little Ice Age" that lasted from about 1350 A.D. to about 1850 A.D. During this 500 year period there was plenty of disease, famine and death, accompanied by hangings or burnings of the "witches" who supposedly brought on these cold times. Huddling humanity together in cold, poorly ventilated spaces with inadequate nourishment created a disease-distributing environment. This was a time when Earth's global temperature averaged only about two degrees Celsius cooler than today!

By contrast, the "Medieval Warm Period" of 1,000 years ago and the "Roman Warm Period" of 2,000 years ago saw humanity flourish and increase in population and well-being. As you will see, this Little Ice Age was nothing but a short interlude in the overall global warmth man has enjoyed for the last 10,000 years.

Fig. 4

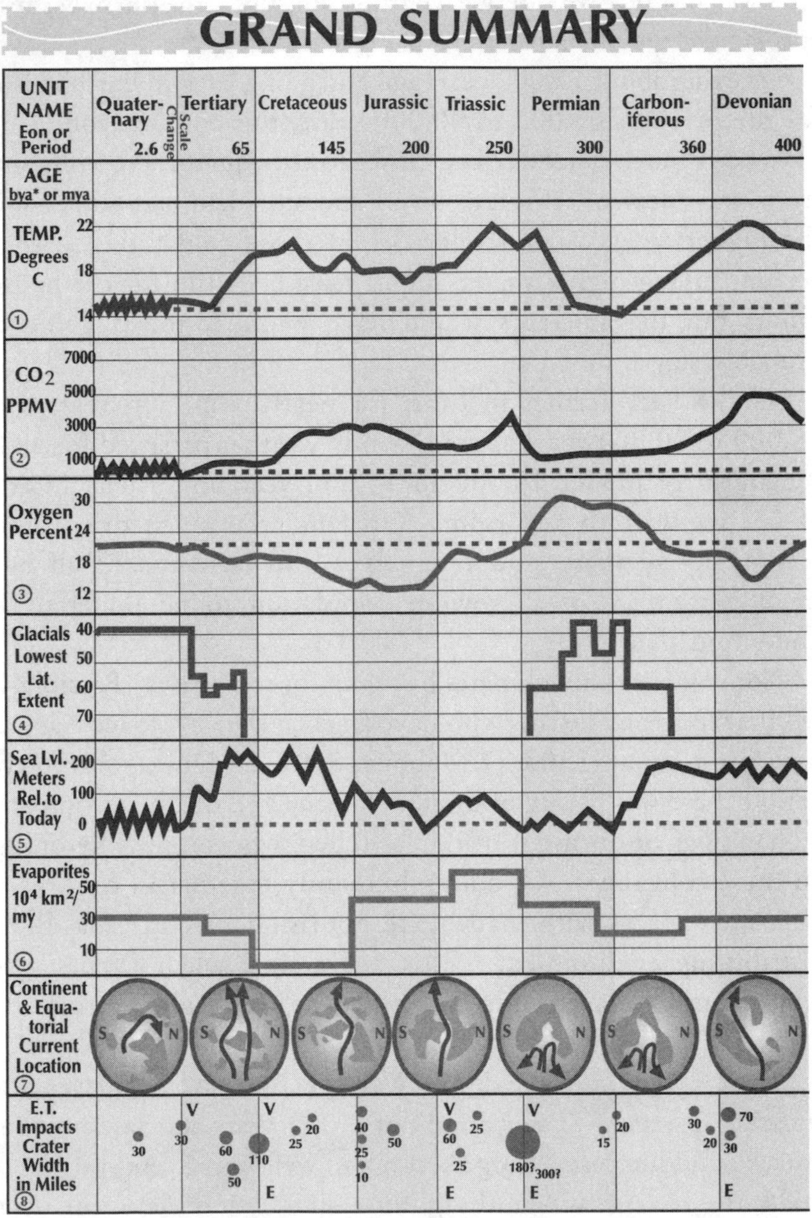
GRAND SUMMARY
UNIT NAME Eon or Period
Quater-nary 2.6
Scale Change
Tertiary 65
Cretaceous 145
Jurassic 200
Triassic 250
Permian 300
Carbon-iferous 360
Devonian 400
AGE bya* or mya
TEMP. Degrees C
22
18
14
①
CO2 PPMV
7000
5000
3000
1000
②
Oxygen Percent
30
24
18
12
③
Glacials Lowest Lat. Extent
40
50
60
70
④
Sea Lvl. Meters Rel.to Today
200
100
0
⑤
Evaporites 10⁴ km²/my
50
30
10
⑥
Continent & Equatorial Current Location
⑦
S
N
E.T. Impacts Crater Width in Miles
⑧
V
E
30
30
60
50
110
25
20
40
25
10
50
60
25
25
180?
300?
15
20
30
20
70
30

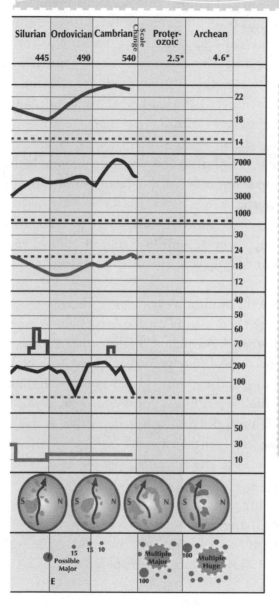

Silurian	Ordovician	Cambrian	Scale Change	Proter-ozoic	Archean	
445	490	540		2.5*	4.6*	

LEGEND

•••••• = Earth's Current Values
bya = Billion Years Ago
mya = Million Years Ago
ppmv = Parts Per Million by Volume
N & S = North and South Pole Locations
↝ = Equatorial Ocean Currents
E = Major Extinction Event
V = Major Volcanic Event

①Adapted after Royer, et al., 2004, GSA Today

②Adapted after Berner, 2004, Oxford University Press

③Adapted after Berner, 2004, Oxford University Press

④Adapted after Crowley, 1998, Oxford University Press

⑤Adapted after Hallum, 1984, Earth Planetary Science 12, & Vail, 1977, AAPG Memoir 26.

⑥Adapted after Gordon, 1975, Journal of Geology 83

⑦Adapted after Scotese, 2005, www.scotese.com

⑧Adapted after University of New Brunswick, 2008

Refer to Chapter Five for discussions of each of the time intervals.

Compiled by
H. Leighton Steward
2009

5

Had our predecessors existed 40 or 50 million years ago, the warmth may have been bearable and the ocean temperatures right for swimming, at least until an ancestor of *Perusarus*, a giant crocodilian creature that survived until only six to nine million years ago in (you guessed it) Peru, came along and gulped our fictitious predecessor down in one sitting if not one bite. The skull of Perusarus measures nearly six feet in length, indicating an alligator - creature probably 40 to 50 feet long and weighing 18 to 22 tons! Compare this with the Tyrannosaurus rex, which weighed in at about 7 to 9 tons. A replica of the skull of this holdover from the age of dinosaurs can be seen at the Audubon Zoo in New Orleans, Louisiana.

During the age of dinosaurs, 65 to 250 million years ago, you would not have been safe on land or in the sea. The mammals that existed were very small and spent most of the daylight hours hiding from their reptilian counterparts.

From 250 million to 325 million years ago, Earth was in the grips of another ice age and from 325 million to 430 million years ago, Earth was much warmer but oxygen content was much lower than today.

How about beyond 430 million years ago? There were no dinosaurs, but neither were there plants on the land. No greenery whatsoever! The period when land plants finally appeared was in only the most recent 10 percent of Earth's history. Imagine a bleak, barren surface with no land-dwelling animals, not even amphibians. Nothing could inhabit a lifeless landmass upon which there was no shade or anything to eat. To put the timing in perspective see Figure 5, a timescale highlighting key events in Earth's history. While you are reviewing Figure 5 also note the time that people have existed on Earth.

Earth was birthed in fire, sometimes dominated by ice and, I truly believe, we are now living in a paradise for our species. We have an abundance of grains from our grasses, nuts and shade from our trees, and on and on. How is our current paradise looking to you now?

Fig. 5

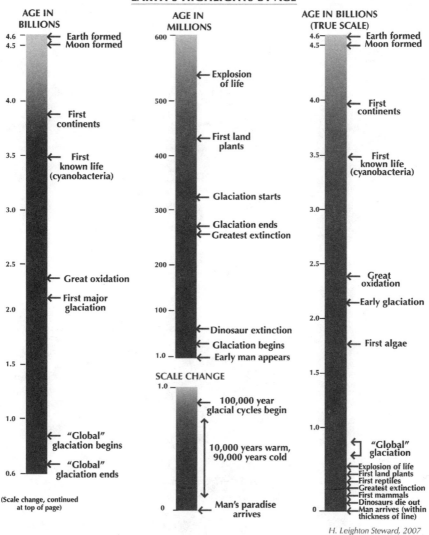

H. Leighton Steward, 2007

7

Chapter Two

THE LIBRARY IS OPEN!

Our ability to study Earth's paleoclimate history has grown rapidly. However, much of Earth's surface has not yet been researched with modern tools. What we do know has come from analyses of only some of the rocks, sediment, ancient soils, minerals, pollen, fossils, corals and even tree rings and leaves. These past climate indicators will be described more thoroughly in Chapter Four. The rocks and sediments that hold the information are vast in quantity and have barely been touched by the types of tools available to researchers today because so many science graduates have gone into the energy, medical, or manufacturing industries. But for those who like meaningful research, scientists entering this field today will have a lifetime of discoveries coming from their probes into the past.

Since tools for what the rock, sediment, and ice records in Earth's "library" of history are telling us regarding past climates have existed for only a few decades, progress certainly will be made in our ability to translate much more from these ancient Earth records. An example of a development that allows scientists to glean more information from Earth's library is the stable isotope analysis of the mineral geothite which has helped us learn more about the carbon dioxide content of the ancient atmosphere (Professors Crayton Yapp and Neil Tabor, Southern Methodist University, personal communication). Geothite is formed in soils today, as well as very old soils, and retains information about the content of the gases from the current and ancient atmospheres as well as Earth's approximate temperatures. Another relatively new tool is found in using an isotope of beryllium (Be^{10}), in addition to an isotope of carbon (C ^{14}) to analyze the intensity and magnetic variations of the sun and cosmic ray influences, which may be very important in unlocking more secrets in why Earth's climate has changed as it has.

The first geologic map, that simply depicted sediment layers of the same age, was made less than 200 years ago. William Smith of England used certain fossils and observations of similar rock types to correlate from one hillside outcrop to the next. He noticed the similarities of certain fossil types in layers of similar sediment layers as he was supervising the excavation of hundreds of miles of canals that traversed the English countryside. This first map, that covered much of the near surface of England, is still surprisingly accurate. Author Simon Winchester has documented Smith's feat in his book, *The Map That Changed The World* (Winchester, 2001). Winchester's book also demonstrates the reluctance of Smith's contemporaries to timely grasp the scientific significance of Smith's observations.

Chapter Four describes what tools are available to help researchers understand what the "books" in Earth's library contain. Hopefully, you'll never drive by a highway road cut or a natural hillside outcrop of rock without realizing how many thousands or millions of years of Earth's history are contained in those rocks and what helpful information may be residing there. Has anyone even sampled or researched that particular "page" (outcrop)? Not many of these sites worldwide have been analyzed with all the tools available today. Fortunately, several sites around the world have been extensively measured, and that is how we have obtained our current level of knowledge.

Describing how we interpret the data held in our worldwide library today is fairly straight-forward, but the integration of the various pieces of data to determine how the climate was affected is very complex. The analyses will only get better as technology continues to advance and as more of the data from Earth's library is studied.

Beyond what happened in the past, the much more difficult questions to answer are why the climate was like it was and what combination of factors (and there are many) caused the climate to change from one state to another.

Scientists have documented the technical data that exist from our very cursory peek into this library which is as large as all of Earth's outer layers, including those layers that exist at the bottom of the current and past oceans, lakes, and marshes. An expansion of the library that is accessible at Earth's surface includes the tens of thousands of cores drilled by the mining and energy industries, many from several miles deep in the Earth. With the geoscientists focusing elsewhere, few have been sampled for paleoclimate implications.

Researchers also have gone beyond earthly bounds and have measured and/or calculated the effects of the sun's varying contributions to Earth's climate as well as the gravitational pulls and tugs on the shape of Earth's orbit around the Sun. Researchers also have uncovered direct evidence of many significant meteor and/or comet impacts that will be described in Chapter Three.

To test whether our interpretations of the causes or drivers of the paleoclimates and the changes that have occurred make good sense, some researchers make models of what may have happened by inputting the best data available. But, as mentioned earlier, where the sampling of the library has been sparse, and it generally has been, confidence in the model's output can be quite low. Thus our need to discover and analyze many more data points in Earth's library so future models will contain all the pertinent data that need to be input to make the models more reliable in forecasting past and future climates.

What does the knowledge of Earth's climate history really do for us? We hope it will allow us to intelligently use our growing array of earthly tools to try to preserve or at least extend this paradise of a climate that is now unusually beneficial to mankind's comfortable existence on Earth. Short of this goal, we hope to be able to forecast the probability and timing of, a future major change so we can be prepared for the expected change.

Figure 6 gives an example of just a few of the growing list of questions that researches have asked as they visited some of the pages in Earth's library.

Fig. 30

Fig. 6

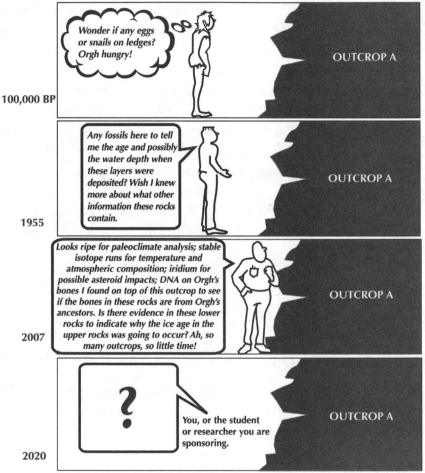

Chapter Three

CLIMATE DRIVERS

I have asked nearly everyone with whom I have conversed since I started writing this book what they think is causing Earth's climate to get warmer. The nearly unanimous answer from the non-scientists is "carbon dioxide" or "man-made carbon dioxide", except from those people who have been studying how the climate really works. Read this chapter and decide for yourself. If you get discouraged trying to remember all of the drivers, just put a paper clip on Figure 7, the one-page summary, and refer back to it as needed. Yes, there are that many drivers!

Sun's Luminosity or Irradiance

The sun provides nearly all of the heat that reaches Earth's atmosphere and surface. A small amount comes to the surface from heat within Earth, such as that which is emitted from volcanoes or sea floor spreading centers. Some of the sun's heat is absorbed at the surface, some is reflected back into space, and some is radiated and retained in the atmosphere by the effect of the greenhouse gases, primarily water vapor, carbon dioxide, and methane. Besides greenhouse gases, there are many drivers that affect or modify how much of and where the sun's heat is retained. As you read on, you will understand how these other drivers come into play regarding global climate and global climate change.

As the Sun's visible energy (light) encounters Earth's atmosphere, most of this energy passes through the atmosphere and strikes the planet's surface. Clouds and dust or other particulates (aerosols) do reflect some of the visible light back into space. Of the dominant amount of the visible light that reaches Earth's surface, nearly half is absorbed, more than a quarter is reflected back into space, and the rest is radiated back as non-visible infrared energy and temporarily "trapped" by the various greenhouses gases (Figure 8). This "trapped" heat helps keep Earth's lower atmosphere warm and Earth's surface from being frozen.

The amount of the Sun's energy that is reflected back to space varies as the reflectivity of the surface changes. For example, snow reflects more visible light and dark soil reflects less visible light, so dark soils absorb more heat and then radiate more infrared heat than snow. The amount of the sun's energy that gets reflected by Earth's surface or atmosphere is referred to as Earth's *albedo*.

Fig. 7

CLIMATE DRIVERS

DRIVER	PRINCIPAL INFLUENCE	IMPACT	COMMENTS
Sun's heat and magnetic variations	Amount of sun's heat and solar shielding variations over time.	Strongest	Amount retained modified by other drivers.
Orbital eccentricity	Determines distance from the sun at any given time.	Strong	Distance variations mean heat variation.
Earth's tilt	Determines Earth's seasons and heat received at high latitudes.	Strong	Additional tilt can affect polar ice melting.
Earth's wobble (precession)	Determines season closest to or farthest from the sun.	Strong	Can be a positive or negative feedback.
GHG water vapor	Strongest GHG. Affects cloudiness, albedo, vegetation, and precipitation volumes.	Strongest of GHGs	Net effect the least understood (predictable) of the GHGs.
GHG carbon dioxide (CO_2)	Captures infrared heat radiated from Earth's surface, reradiates some heat.	Strong at low saturation	Greenhouse effect non-linear. Usually follows temperature changes.
GHG methane CH_4	Captures infrared heat radiated from Earth's surface, reradiates some heat.	Moderate (low volume in atmosphere)	Generated in wetlands, by industries and some animals.
Ocean currents	Distributes heat around Earth. Can change patterns quickly or very slowly.	Very strong	Largest reservoir of surface heat. Affects precipitation.
Plate tectonics (seafloor spreading)	Causes volcanism, CO_2 and sulfate particle input, subduction and mountain building.	Strong long-term	Affects ocean currents, sea levels, and CO_2 volumes.
Location of continents	Affects major ocean currents and heat and moisture to poles.	Strong to weak	Land over poles promotes more glaciation.
Elevation of land masses	High elevations increase chemical weathering (CO_2 removal).	Moderate	Also affects regional climates, monsoons and locations of deserts.
Chemical weathering	Affects CO_2 removal and carbon sequestration.	Moderate long-term	Little short-term effect on climate.
Volcanism	Constant source of CO_2, sulphate particles, and short-term soot.	Moderate to strong short-term	Provides great lava and ash layers for age dating.
Extraterrestrial impacts	Immediate fires, then cold for 1 to 5 years.	Strong very short-term	Can create ocean and atmospheric toxicity.
Albedo	Determines how much solar heat is reflected or retained.	Moderate to strong	Affected by many drivers, constantly changing.
Fauna & flora (animal life and vegetation)	Affects albedo and oxygen, CO_2 and methane content of atmosphere.	Moderate	Abundance and type of flora track temperature, CO_2, and moisture changes.
Atmospheric circulation	Distributes heat and moisture and affects upper ocean current patterns.	Moderate	Distributes nutrients to oceans, affecting sea life abundance and carbon sequestration.
Cosmic rays	Suggested they create particulates that seed low level clouds (cooling).	To be determined	More research needed to verify impact magnitude.

H. Leighton Steward, 2007

Fig. 8

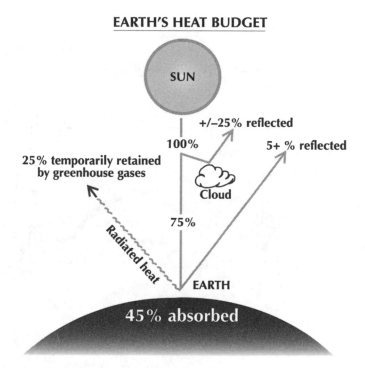

EARTH'S HEAT BUDGET

This diagram shows a very simplified approximation of the retention or reflection of the sun's incoming energy. About 70% is retained and 30% is lost back to space.

H. Leighton Steward, 2007

Earth's albedo changes just as your albedo changes if you wear a white shirt in the sunshine rather than a dark shirt. You are cooler in the high-albedo white shirt. Variations in Earth's albedo are very important in affecting global climate. Albedo will be discussed more thoroughly later in the chapter.

The energy output of the sun does vary. The best known example is the sun's cycle that recurs approximately every 11 years - the 11-year sunspot cycle. At the time of maximum sunspots on the surface of the sun, the heat released from the sun increases somewhat as a result of the presence of these dark spots that are surrounded by intense haloes (faculae) that emit an extra measure of light (heat) that is sent toward Earth. There is also a variation in how long the sunspot cycles last. To see how well the sunspot cycle length in years correlates with Earth's temperature variations, see Figure 9. The shorter the cycles, the higher the temperature.

Fig. 9

SUNSPOT CYCLES & TEMPERATURE TRACK

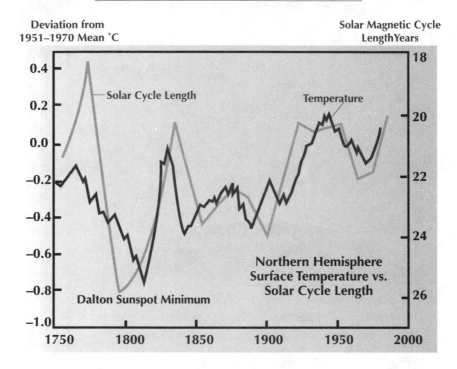

Deviation from 1951–1970 Mean °C

Solar Magnetic Cycle Length Years

Solar magnetic cycle length and Earth's temperature: a strong correlation.

Adapted from A. Pekarek, 2001

There are other times, less well documented, when the variation in the sun's energy output is greater or lesser and it is not fully known what drives some of these longer solar cycles. It may be controlled by the changes of magnetic activity levels within the sun. We do know that solar radiation can vary beyond the common 11-year cycles. There was a 70 year long period (1645 to 1715 A.D.), the Maunder Minimum, that coincided with the coldest part of the Little Ice Age during which time sunspot activity decreased and the sun's delivery of heat to Earth diminished about one half of one percent and Earth cooled. Solar activity, represented by sunspot cycle length and magnetic variations, indicates that solar activity, including outpouring of solar wind, has been high for most of the last century and Earth experienced overall warming.

Solar intensity variation also is believed to occur in 87 and 210-year cycles (Gerhard, 2007) and the occasional coming-in phase of the cycles is believed by some scientists to cause the observed short-term warm cycles that occur about every 1,500 years (Singer and Avery, 2007). This recurring cycle, regardless of origin or duration, has been well documented by Gerald Bond et al., 1997. That this 1,500-year cycle has been very well documented but has left so many capable scientists scratching their heads as to its cause, is just another example of the complex array of factors that determine global climate change.

A very long-term trend has seen the sun's overall heat output increase by approximately 30 percent in the last 4.6 billion years. This overall change is so slow we won't have to worry about it in the next century or millennium. It is interesting to speculate, however, that if the sun's output was not 30 percent higher than in the past, our paradise (and ourselves) probably would not exist, and our already cooler-than-average Earth would be in a frozen, snowball state. To see that Earth is now cooler than average, despite "global warming", refer again to Figure 4, page 4.

To see how much of the sun's energy strikes an area of Earth's surface near the equator versus that same amount of energy striking Earth near the poles, see Figure 10, page 19. A low angle of contact near the poles will spread the same amount of energy over a much greater land surface, which means much less heat is received and retained for a given area of land or water. Other factors are important in determining the amount of, and where, when, and how the sun's heat reaches Earth, as well as how much is retained.

Orbital and Rotational Factors

Very important factors in determining long-term climate changes are the shape of Earth's orbit around the sun and the changes in what part of Earth is facing the sun at any given time. Following are the effects of the various orbital and rotational factors at play:

Orbital Eccentricity

If Earth's orbit around the sun was perfectly circular, which it sometimes nearly is, Earth would receive the same amount of solar heat each year for its entire 365-day orbit around the sun. If, however, the orbital path was more elliptical, Earth sometimes would be closer to the sun during part of the year and sometimes farther away and would not receive a constant level of solar heating. This is exactly what happens. Earth's orbit gradually changes from nearly circular to somewhat elliptical and back over a period of approximately 100,000 years (Figure 11, page 20). Earth is currently in a more circular-shaped orbit, although it is always somewhat elliptical.

Today the variation in Earth's maximum and minimum distance from the sun in its 365-day orbit is as much as three million miles. While that may not seem like much in an average distance of 93 million miles, it amounts to a modest difference in heat received and apparently has been strong enough in its extremes that, when combined with what part of Earth receives heat and when it is received (i.e. summer or winter), it has caused the last eight or ten glacial - interglacial cycles that have occurred over these 100,000 year intervals. Studies of ice cores reveal that the orbital and rotational factors have been primary drivers of Earth's major climate changes for the last several hundred thousand years and undoubtedly for eons past to a greater or lesser extent.

To observe the astonishing regularity of the last four glacial - interglacial cycles, which have been very well documented through the studies of ice cores from Antarctica, see Figure 12, page 21. All but the last 10,000 years, the current warm period, occurred without any influence by the human race.

As Paul Mayewski and Frank White pointed out in their book *The Ice Chronicles*, (Mayewski, P.A. and White, F., 2002) the more order we see in nature, the better chance we have of predicting the future. I like their observation. However, I also know that some wise soul said, "Nothing ever stays the same." One day, the influence of a combination of climate drivers will converge and take us out of this recent rhythmic cycle and onto a new one. For our future generations, let's hope it morphs into an extended period of paradise! Unfortunately, with the ever-changing moving parts such as orbital variations and solar intensity changes, I doubt the permanent paradise will ever occur unless mankind achieves power beyond my current imagination. But then, look at what man has thought would never occur but has happened - lights, flight, atomic power, moon walks, landing on asteroids, a tiny injection to prevent horrible diseases, and talking to someone anywhere on Earth or in space from a tiny handheld devise. High-tech is in it infancy; some of us can remember no air conditioning or television! What's next? Anti gravity? Warp speed? Time travel? High tech is not "20 seconds" old. Why preclude controlling the climate? But let's get back to today.

Fig. 10

SOLAR HEAT DISTRIBUTION BY LATITUDE

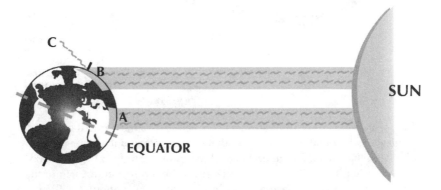

For a given medium, say water, more of the sun's heat is absorbed per square meter near the equator (A) than near the poles (B) because an equal amount of the sun's energy is spread out over a greater surface area (in effect, diluted) near the poles. Also, the lower angle of contact with the surface results in still more of the sun's rays being reflected back into space (C).

H. Leighton Steward, 2007

Fig. 11

SHAPE OF EARTH'S ORBIT

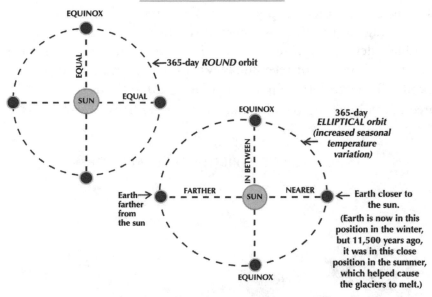

There is approximately 100,000 years between Earth's most round and most elliptical orbit around the sun. This is about the same amount of time as there is between the glacial and interglacial climates on Earth for at least the last 800,000 years. The difference in the closest orbital position of Earth to the sun today and in Earth's farthest position from the sun is about 3 million miles, which affects the amount of energy that strikes Earth's atmosphere.

H. Leighton Steward, 2007

Fig. 12

THE LAST FOUR GLACIAL CYCLES

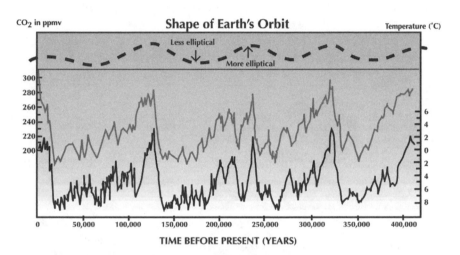

Antarctic temperature and CO_2 history from ice core analyses. Note the amazing rhythmical similarities of the four cycles, which indicate the very strong solar-orbital influences on Earth's climate. Since Earth's CO_2 levels do not drive the solar-orbital cycles, you can see why many scientists doubt the currently popular "CO_2 CAUSES global warming" argument. On the other hand, an increase in CO_2 does provide some positive feedback (support) to a warming Earth, but the magnitude of that effect is still being debated.

Adapted from Pettit, et al., 1999

Earth's Tilt Axis

Earth's axis of rotation is aligned with the north and south poles and is actually tilted relative to Earth's orbital plane around the sun (Figure 13). This tilt is what gives us the seasons. If the equator was lined up directly with the plane of our 365-day orbital path around the sun, the sun's rays would fall on exactly the same latitude with the same intensity for the entire year and there would be no winter, spring, summer, or fall. Many astrophysicists think that a Mars-sized body that collided with Earth and formed the moon changed or tilted Earth's axis of rotation and gave Earth's its seasons.

Fig. 13

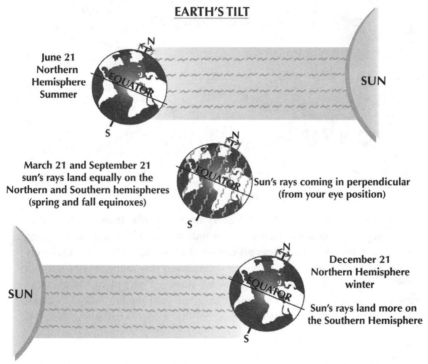

EARTH'S TILT

June 21 Northern Hemisphere Summer

SUN

March 21 and September 21 sun's rays land equally on the Northern and Southern hemispheres (spring and fall equinoxes)

Sun's rays coming in perpendicular (from your eye position)

SUN

December 21 Northern Hemisphere winter

Sun's rays land more on the Southern Hemisphere

Earth's tilt ranges from maximum to minimum tilt every 41,000 years. Scientists believe tilt is important because prior to about 1,000,000 years ago, Earth's glacial-interglacial cycles were on 41,000-year intervals.

H. Leighton Steward, 2007

The tilt angle is currently about 23.5 degrees, but it varies over time, from 21.8 to 24.4 degrees, in about a 41,000-year period. This change in tilt does have an effect on the climate by increasing or decreasing how much of the sun's energy lands on Earth's poles. For instance, the greater the angle of tilt, the more of the sun's energy shines down on the poles, which, in the summer, can contribute to additional melting of ice on the pole tilted toward the sun at that time. Obviously, the opposite pole would be a little colder in its winter, because more of that pole's ice or land would be hidden from the sun's rays. If other drivers are collectively exerting a warming or cooling influence on Earth, the tilt angle could be more important in helping to cause or in dampening the magnitude of a climate shift.

An example of the possible major influence of changes in Earth's angle of tilt lies in the fact that beyond the last one million years, Earth's glacial periods occurred at approximately 41,000 year intervals; the same timing as one complete tilt cycle.

Earth's Wobble

Like the slowing of a spinning top, Earth also wobbles with one full wobble every 19,000 to 23,000 years. This wobble causes the seasons to change depending on whether the time when the sun is closest to, or farthest from, Earth falls in summer or in winter (Figure 14, page 24). Again, this can have an effect of increasing or decreasing what spot on Earth receives the most or least heat and in what season. Since there is more low-albedo (less reflective) water in the Southern Hemisphere and more highly reflective landmass in the Northern Hemisphere, when and where the sun is delivering its heat can exert a significant influence on heat absorption and global climate. Global climate does not change in lock-step due to these differences.

These variations in the length and intensity of heating of different parts of Earth's geography clearly have strong effects on the climate. For instance, 11,500 years ago, the time when Earth was the closest to the sun in Earth's eccentric orbit occurred in July, which caused very hot and lengthy summers and, combined with orbital and tilt influences, resulted in the melting of most of Earth's ice cover from the last ice age. While the winters were very cold, the ice and snow could not accumulate fast enough to keep up with the intense summer melting (again, see Figure 11, page 20) and Earth was temporarily freed from the grips of that ice age and transformed into the current period of paradise. Astronomer's ability to reconstruct ancient orbital positions has helped us understand the global climate changing relationships.

Fig. 14

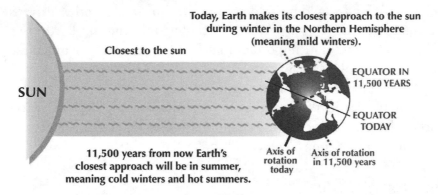

EARTH'S WOBBLE (AXIAL PRECESSION)

Today, Earth makes its closest approach to the sun during winter in the Northern Hemisphere (meaning mild winters).

Closest to the sun

SUN

EQUATOR IN 11,500 YEARS

EQUATOR TODAY

11,500 years from now Earth's closest approach will be in summer, meaning cold winters and hot summers.

Axis of rotation today

Axis of rotation in 11,500 years

As Earth wobbles, the season when Earth is closest to, or farthest from, the sun in Earth's elliptical orbit changes (precesses). Today, Earth is closest to the sun in the winter. In 11,500 years, Earth will be closest to the sun in the summer. Since the average albedo (reflectivity) is different in the two hemispheres (more mid-latitude land in the Northern Hemisphere), this also can have an impact on how much heat is absorbed by Earth for the winter (or summer) season. Will everything be exactly the same in a full cycle (23,000 years)? No, because the orbital eccentricity (100,000-year cycle) and Earth's tilt (41,000-year cycle) will be in a different alignment. Wobble is just another of the ever-changing orbital orientations that helps determine Earth's constantly changing climate.

H. Leighton Steward, 2007

Orbital Summary

Any of the orbital, tilt or wobble variations can have an effect on Earth's climate. Since they complete one full cycle of change or influence over differing periods of time, 100,000, 41,000, or 19,000 to 23,000 years, you can easily imagine that at any particular time they may reinforce each other or, in a different alignment, tend to cancel each other out.

To have all of the orbital factors maxed out toward a colder effect or maxed out toward a warmer effect would be akin to having a "perfect orbital storm" influence on climate at that particular time and push it very strongly in the direction of a colder or warmer climate. Fortunately, since astronomers can calculate when the various alignments will occur that will favor warming or cooling, this will help our distant descendants prepared for such changes.

For the geological near term, hundreds to a few thousand years, the effect of the orbital factors and resulting variations in solar intensity would suggest an influence toward a somewhat cooler Earth than our paradise of the last 10,000 to 12,000 years. This is one reason to sequester (store), in a retrievable place, the greenhouse gases we create from the use of fossil fuels. If Earth begins to cool, we will be able to reach for our stored blanket of CO_2, however modest its effect, to help us stay a little warmer or to increase food production from the critically important greenhouses.

Variations in solar magnetism and solar wind also appear to influence Earth's climate and will be addressed later in the discussion of cosmic rays.

Greenhouse Gases

Hooray for greenhouse gases! Greenhouse gases are very important in retaining the heat radiated from Earth's surface. Unlike other gases present in Earth's atmosphere, such as nitrogen (78 percent) and oxygen (21 percent), which trap almost none of the sun's heat, the greenhouse gases trap non-visible, infrared heat that is radiated back into the atmosphere from Earth's surface and thus helps warm the atmosphere. Without them Earth would be frozen, since most of the sun's incoming energy simply would be reflected and radiated back into space. We don't, however, want too much effect from the greenhouse gas content in the atmosphere, just as we don't want too little of it. Mankind's paradise for the 10,000 years has had just about the right amount of the combination of these gases to help keep the near-surface temperature at an average of 59 degrees Fahrenheit.

Humans have some tolerance for variations in the principal greenhouse gases, possibly much more than we realize. For instance, humans can live in good health at sea level, where the atmospheric carbon dioxide content averages around 0.038 percent, and methane about 0.00002 percent and the oxygen level (not a greenhouse gas) averages around 21.0 percent. Humans can even survive at 15,000 feet where the respective levels are about the same but the total amount is less because of the overall reduction of all gases caused by the altitude and thinness of the atmosphere. We are a lot less tolerant of changes in oxygen and methane (CH_4) than changes in water vapor or CO_2. The two other greenhouse gases, nitrous oxide (N_2O) and ozone (O_3) have much less effect on Earth's climate, even at sea level, because they are present in such small amounts.

Water Vapor (H2O)

Water vapor is by far the most dominant of the greenhouse gases, accounting for as much as 95 percent of them. (Figure 15) It occurs in forms and places that can contribute to either heating or cooling on a global or regional scale. High, bright cloud layers made up of ice crystals temporarily "trap" some heat and thus contribute to high-altitude warming, whereas dense, low-level clouds reflect some of the sun's incoming ultraviolet heat. This has a cooling effect by reflecting sunlight from the low cloud's bright tops. These low-level clouds can also temporarily "trap" some of Earth's radiated, infrared heat. This ability to trap and reradiate heat is most noticeable on cloudy nights when the temperature stays much warmer than on cloudless nights, when Earth's radiated heat can escape to space. When the clouds build up to tremendous heights, some heat energy is lost to space and results in cooling. The answer to the question *"Does water vapor cool or heat the Earth?"* is *"Well, it depends."* Atmospheric water vapor and cloud formation are other areas ripe for research, because they can have large temperature effects, and their net effect is still not well understood.

The average amount of water vapor in Earth's atmosphere today ranges from near zero to four percent. The general content of water vapor in the atmosphere can be inferred by a well-known law of physics: warm air can hold more water vapor than can cold air. Thus, the cold episodes are generally drier (and windier), and the warm periods are generally more moist (with lower average wind velocities). Analyses of ice cores and worldwide sediment cores that cover multiple cold and warm cycles have confirmed this generalization. I know of no scientist who disagrees because the ice-core data that show huge amounts of dust being deposited during the cold periods are very convincing.

Fig. 15

SOURCES OF GREENHOUSE GASES

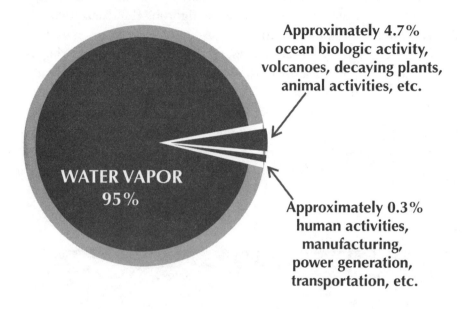

**Approximately 4.7%
ocean biologic activity,
volcanoes, decaying plants,
animal activities, etc.**

**WATER VAPOR
95%**

**Approximately 0.3%
human activities,
manufacturing,
power generation,
transportation, etc.**

Modified after Hieb and Hieb, 2003

Carbon Dioxide (CO2)

Carbon dioxide temporarily absorbs heat in Earth's atmosphere by capturing some of the non-visible, infrared energy that is reflected back toward space. Simplistically, one would think that the more CO_2 in the atmosphere, the more infrared heat is captured in the atmosphere. However, **CO2's capacity to capture additional heat does weaken as CO2 levels continue to rise.** Exactly how much more or less heat is captured and retained by various levels of CO_2 and whether CO_2 is a principal driver of global climate or is simply a greenhouse gas that may provide a modest positive feedback (support) for heat retention, is at the heart of the debate currently being waged about global warming. After considering the detailed analyses of multiple ice cores and calculations of CO_2's logarithmic (rapid) decline in its capability to trap additional heat, I believe that if CO_2 is having an impact on the climate it is likely quite modest (Figures 16 and 17). That CO_2's capacity to trap additional heat declines logarithmically has been cited by many scientists around the world including MIT Professor Richard Lindzen who says it is documented in text books on radiative heat transfer and that the scientific evidence is rigorous (Lindzen, Richard, 2009, personal communication).

However, even if CO_2 at its present atmospheric concentration is not contributing significantly to the current warming trend but is only a weak positive feedback factor, CO_2 is still an important greenhouse gas that, particularly at low concentration in the atmosphere, provides a portion of the overall greenhouse effect that helps keep Earth from being a frozen planet.

Fig. 16

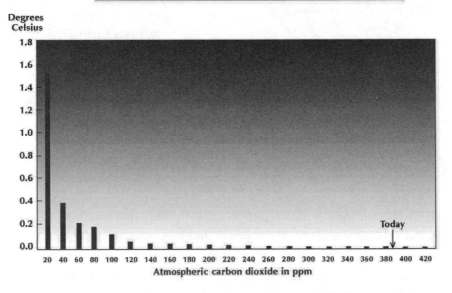

THE WARMING EFFECT OF ATMOSPHERIC CARBON DIOXIDE

Atmospheric carbon dioxide in ppm

David Archibald's chart of the logarithmic decline in CO_2's capacity to trap heat radiated from Earth's surface is excellent in its simplicity. Using the 20-year real time research on Earth's sensitivity to changes in atmospheric CO_2 concentrations (Idso, 2001), Archibald calculates only a 0.2 degree Celsius rise in temperature by the time CO_2 concentrations have reached 620 ppm in the 22nd century.

Modified after I. David Archibald, 2007

Fig. 17

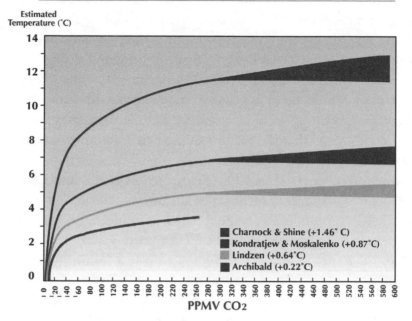

ESTIMATED CLEAR SKY GREENHOUSE EFFECT FROM DOUBLING CO2

These projections support Archibald's hypothesis on major CO2 effect occurring at very low levels of CO2. The variations are caused by the scientists assuming different sensitivities of Earth to CO2. Archibald used Idso's empirical value (Idso, S., 2001).

Modified after S. Milloy, 2007

Before someone tries to get you to think of CO_2 as a pollutant, also consider that it is absolutely critical to supplying the entire spectrum of plants with their required energy to proliferate and start the food chain for nearly all forms of life on Earth. The plants, in return, give back oxygen. That generation of oxygen is the only reason humans can exist on Earth. See Figure 18 that demonstrates the **additional plant growth from doubling carbon dioxide and, despite the impressive additional growth, the plants require less water.**

The CO_2 level has varied dramatically in the last 500 million years, ranging from the geologically recent glacial period lows of approximately 180 *ppm*, to the pre-industrial level of 280 *ppm* (today we are at 380 *ppm*), to a time around 500 million years ago when the level was approximately 7,000 *ppm*, or 18 times higher than today. Global warming **skeptics cite the fact that glaciations have occurred during periods with high CO2**

levels as evidence that CO2 levels were certainly not the drivers of the climate changes and that the predicted logarithmic decline of CO2's ability to trap heat is correct. This latter observation allows us to understand more easily how glaciations and high levels of CO_2 can coexist. It also explains the reason temperature trends and CO_2 trends do not move in lock-step. Again see Figure 4, page 4, for an illustration of the generalized CO_2 and temperature variations over time. While there is a gross correlation with temperature changes over the long-term, detailed studies over somewhat shorter time periods show that the relationship often breaks down (Figures 19 and 20). The fact that there is so much more water vapor in the lower atmosphere than CO_2 (Figure 15, page 27) and that the principal infrared heat absorption bands of CO_2 are almost entirely overlain by those of water vapor, helps us understand the much greater heat-trapping greenhouse effect of water vapor.

Consider this; at sunset in the desert in low humidity and the current 380 *ppm* of CO_2, the ultra hot temperature drops to a chilly level rapidly due to heat easily escaping through the atmosphere. In New Orleans where the humidity is very high and the CO_2 level also 380 *ppm*, the temperature drops very slowly at sunset due to the increased water vapor trapping heat in the lower atmosphere. If CO_2 was such a powerful greenhouse gas, wouldn't it trap the heat escaping in the desert?

Sources of CO_2 include volcanoes, mid-ocean spreading centers, fires, animal respiration, warming of the oceans, cosmic impacts, and combustion of the various hydrocarbons that we use as fuel.

Sinks, or factors that remove CO_2 from the atmosphere, are plant life on the continents or algae in the oceans through photosynthesis, animal life in the form of mega to micro sea shells (including reef-building organisms), calcium carbonate rocks including limestones, lime mudstones, and chalks, burial of plant and animal life that forms coal, oil, and gas, absorption as oceans cool, and chemical weathering processes (to be discussed later). Most of the carbon that is not in the atmosphere or in living things is stored in the various carbon-bearing rocks that I just mentioned.

Fig. 18

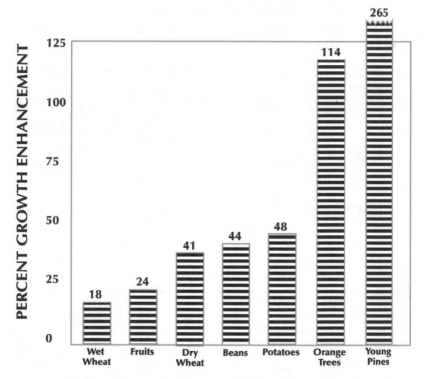

GROWTH ENHANCEMENT FROM 300ppm INCREASE IN C02 *+

265

114

48

44

41

24

18

PERCENT GROWTH ENHANCEMENT

125 — 100 — 75 — 50 — 25 — 0

Wet Wheat | Fruits | Dry Wheat | Beans | Potatoes | Orange Trees | Young Pines

* Greenhouse operators use 1,000 ppm for optimal growth.
+ Our Supreme Court Says CO^2 can be classified a pollutant!

Despite this dramatic increase in growth, the higher CO_2 content causes the plants to need LESS water! Most of the plants on Earth today evolved when atmospheric CO_2 levels were higher so they respond well to the additional CO_2.

H. Leighton Steward after Robinson et al., 2007 and Idso, 2007

Fig. 19

NORTHERN HEMISPHERE TEMPERATURE
VS. SOLAR IRRADIANCE

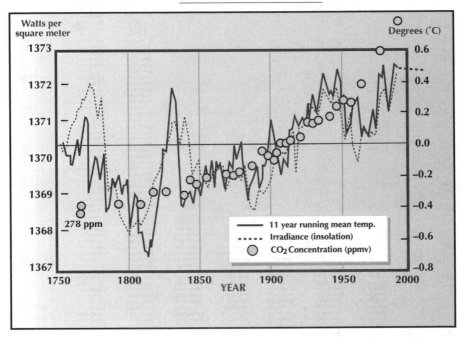

CO2 and temperature don't co-vary. Solar energy and temperature do co-vary.

Adapted from Hoyt & Shatton, 1997
and Keelig & Whorf, 1996

Fig. 20

21st CENTURY CLIMATE FROM THE U.K. HADLEY CLIMATE CENTER

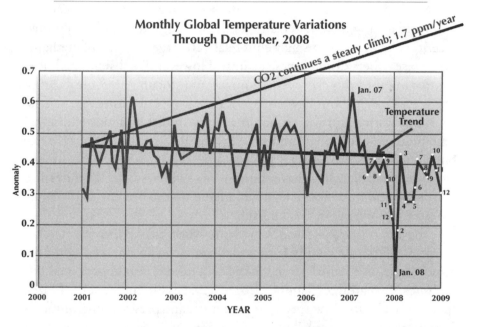

Monthly Global Temperature Variations Through December, 2008

This gradually declining temperature trend line (see arrow) was calculated through mid-2007. It has clearly turned more downward since then.

Compiled by H. Leighton Steward

Methane (CH4)

Methane is what is commonly referred to as natural gas. Methane is another, even more powerful greenhouse gas that can trap more than 20 times more radiated infrared heat than an equal volume of carbon dioxide. It does, however, remain in the atmosphere only about 10 years and absent a constant resupply, there is not much potential for a long-term buildup of methane. Fortunately, for those who fear any additional global warming effect, the amount of methane in the atmosphere today is 200 times lower by volume than current CO_2 levels.

Methane levels increase as swampy areas on Earth increase, because plant life rotting in water creates methane. Also, it has been reported recently that living plants may emit relatively large volumes of methane; other researchers dispute these claims. However, if substantiated, isn't it amazing that this fact wasn't known and documented until 2006? See what I mean about new discoveries awaiting our future researchers? As Earth warms and the air becomes more moist and aids additional precipitation, additional swampy areas should develop. Extensive rice farming over the last several thousand years has caused a measurable rise in methane levels, as have the emissions of livestock. Termites also emit significant quantities of the methane that is present in the small amount of methane in the atmosphere.

The tundra region releases methane when the environment turns warmer and the previously frozen methane deposits are released to the atmosphere. Methane is stored in the cold tundra peat bogs as methane hydrates or clathrates (simply frozen crystals of methane). Frozen methane is also found at specific temperature and pressure environments in sediments at or just below the sea floor. Sub sea methane volumes can be huge and their sudden release can have a warming influence, albeit temporary, on global climate. These deposits can be released by submarine landslides, submarine volcanism, or even by very large impacts from space that directly destabilize the temperatures and pressure environment that is harboring the methane.

Ozone (O3)

Ozone is a trace gas present in very small amounts (parts per billion) in Earth's atmosphere. Ozone's most important role relative to human life is that it absorbs incoming ultraviolet light and also protects us from skin cancer and damage to our eyes.

Beyond absorbing the ultraviolet rays, it does cause a minor amount of warming at the near surface and cause some warming in the stratosphere by amplifying changes caused by increases in solar irradiance. Ozone molecules, formed by solar ultraviolet radiation striking oxygen molecules, absorb some of the solar radiation and heat the upper atmosphere somewhat.

Nitrous Oxide (N2O)

Nitrous oxide is also a trace gas present in very small quantities (parts per billion), and while it captures some radiated heat coming from Earth's surface, its contribution appears to be very small and likely does not drive global climate change.

Greenhouse Commentary

Now that we know what the three chief greenhouse gases are and what they do, you may wonder why all the fuss over a little variation in carbon dioxide content, especially considering that some **well-respected scientists** like H. M. Fischer et al., 1999, U. Siegenthaler et al., 2005, N. Callion et al., 2003, Monnin, et al., 2005 and Muddlesee et al., 2001, **have determined that the changes in CO2 levels during recent glacial cycles FOLLOWED, not CAUSED, the temperature changes** and the lag in the CO_2 response was several hundred years as determined by all of the researchers. An example of the temperature and CO_2 relationship is demonstrated in Figure 21. By the way, ditto for methane increases FOLLOWING the warming and cooling. I believe that prior to Fischer, et al.'s discovery of CO_2 lagging the temperature changes, so many scientists and public figures had been convinced of the dominant role of CO_2 as a climate driver and had so convinced the general public (as well as myself) that there has been a reluctance on the part of the large majority to admit that the CO_2 role is much less than previously thought. Our system does not reward someone for being wrong.

Fig. 21

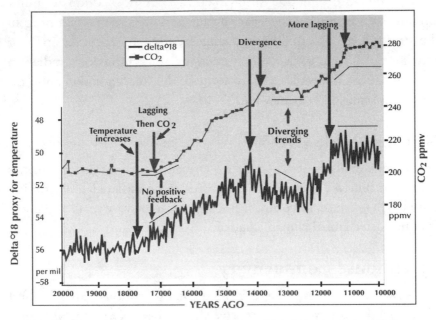

ICE CORE ANALYSES–ANTARCTICA

This detailed analysis clearly shows the several hundred year lag of CO2's response to a prior temperature change. A cause does not follow an effect. It also illustrates that CO2 and temperature do not always co-vary and there is no indication of positive feedback.

Concordia (Dome C) Analyses after L.E.Monnin, et al., 2005

Another climate scientist, Dr. Sherwood B. Idso, (2001), points out that the warmer than "normal" *Medieval Warm Period* and the cooler than "normal" *Little Ice Age* both occurred in the last 1,000 years accompanied by no significant change in CO_2 levels as measured in ice cores. Other scientists have said ice core measurements from depths greater than 200 meters are not accurate. Regardless, CO_2 changes clearly seem to follow temperature changes. Idso also performed several real-world observations over a 20 year period while he was working at the United States Water Conservation Laboratory in Phoenix, Arizona. The results of this research indicated that **the climate of Earth is not very sensitive to further increases in CO2** and his best estimate of doubling CO_2 in the atmosphere from 300 to 600 *ppm* would probably be only about 0.4 of a degree Celsius.

The crux of today's concern is that no one wants the temperature to change much in either direction. We know greenhouse gases temporarily trap some heat and, if increased enough, can cause or help the atmosphere to warm by re-radiating some of that captured heat. A reduction in greenhouse gases can allow Earth to lose some heat to space and help cause cooling unless other climate drivers, as many scientists believe, overpower the greenhouse effect. A significant change in either direction and our climate could move out of the paradise zone. While some scientists are still struggling with whether increases of CO_2 beyond the current levels have significant warming effect, let's not continue to add CO_2 at such an unprecedented rate, because we still do not know the precise answers as to exactly how much it may cause some other effect such as making the oceans less alkaline. Moving toward acidity could possibly affect the development of the shells of some of the life in the ocean that are at the beginning of the Ocean's food chain. Other scientists such as Dr. Tom Segalstad of Norway (personal communication) says that there are such large volumes of natural buffers to acidification in ocean water, that the water becoming acidic can not happen.

While it is probable that some warming will occur with the input of more of the greenhouse gases, we don't exactly know what magnitude of change it will take to push the temperature out of the paradise zone. Many people have decided that any man-made additions are bad and have endorsed dramatic, near-term decreases in greenhouse gas emissions. While I can understand this view, I endorse a more phased reduction in greenhouse gas emissions <u>at a rate</u> that will not cause a significant reduction in mankind's standard of living and the amount of money we spend on climate research, medical research, improving appliance efficiencies, developing alternate energy sources that make sense, and the like. Currently, however, there is a question of whether CO_2 can be sequestered at a reasonable price or in a secure reservoir from which the extremely soluble CO_2 cannot ultimately escape. This is another area where more research dollars should be provided because, if the technology becomes available to sequester CO_2, a storehouse of CO_2 could prove useful in greenhouses when Earth begins to cool.

Earth's orbital orientation is now slowly changing from a configuration that promoted recent interglacial warming to one that will favor global cooling. Many scientists, including Professor William Ruddiman at the University of Virginia, (Ruddiman, William F. 2001) have suggested that the only thing that may have kept us out of the next glacial cycle, has been man's agricultural success over the last 8,000 to 10,000 years, which has generated additional methane (a stronger greenhouse gas than CO_2) and the release of CO_2 through the burning of fossil fuels (coal, oil, and natural gas).

So, while most people are trying to minimize the current greenhouse gas buildup, let's boost our climate-change research so we will know more about how and when to use any means at our disposal to temper the sudden and significant climate changes such as those that have occurred over and over and over again in Earth's natural, pre-human history.

Of general interest is a non-greenhouse gas, oxygen. In paleohistory, long, long ago, oxygen (O_2) levels were at one time zero. Earth first became oxygenated significantly about 2.4 billion years ago. Oxygen levels later rose, about 275 million years ago, to make up approximately 30 percent of the atmosphere and fell at another time, about 200 million years ago, to approximately 12 percent of the atmosphere. When oxygen levels were very high, wildfires abounded and when the levels got very low, many species went extinct. Oxygen levels today are about 21 percent, which is ideal for us and the other inhabitants of the planet. When considering the role of CO_2 in feeding the plants, remember again that plants provide us with the oxygen we breathe.

Ocean Currents

Approximately 70 percent of Earth's surface is covered with water. On the average, water absorbs much more and reflects much less of the sun's incoming heat than does land. The oceans hold a tremendous amount of heat and are very important in affecting Earth's climate.

The oceans give off heat when the water is warmer than the overlying atmosphere which is why many coastal areas have more temperate climates than land located farther from the coast. Conversely, if the water is cooler than the atmosphere, it will tend to cool the otherwise hot atmosphere and also temper the adjacent coastal climates.

With effects from varying wind patterns, gravitational effects on the tides, differences in water salinity, and differences in temperature, the oceans develop currents that transport these warmer or cooler waters around the globe. A significant change in these current patterns can greatly influence Earth's climate in a regional, if not global, manner.

What lets us predict when the world's major ocean currents will change patterns? At this moment, the answer is much more than I or anyone can tell you, because this remains an understudied, under-measured area of science. One obvious influence is the changing locations of the continents, but ocean currents change even when the continents have hardly moved. The data gathering is increasing rapidly and this is another area ripe for discovery.

Besides simply affecting the temperature, ocean currents can affect the amount of moisture in the air. Warmer air can hold more moisture than colder air. So, if you want more rain or snow, having more moisture in the air is very helpful.

Since colder air doesn't hold as much moisture as warm air, if a warm current should flow up into the northern latitudes and provide a little more moist, overlying air, you should expect to get more rain or snowfall than normal. This is what may have happened when the warm Gulf Stream began to flow to the North Atlantic when the opening to the Pacific Ocean closed in Panama three million years ago. In somewhat cooler times, this moist air that overlies the warm Gulf Stream has allowed the glaciers and ice sheets to grow more rapidly and amplify the ice buildups during the glacial cycles.

In contrast, cold-water current upwelling adjacent to a mid-latitude shoreline can create a colder, low-water-vapor atmosphere and cause very dry conditions that result in little rain or snow. This is what is happening just onshore in the Atacama Desert (the driest place on Earth) on the west coast of South America, even though the land area is directly adjacent to the Pacific Ocean.

Over time, usually a long interval such as tens of thousands or millions of years, the major ocean currents do change their patter of flow, and in some instances, these changes cause major climate changes. Again, let's look at the Gulf Stream as an example of how a change in current flow or position can change or help change the climate. As pointed out by Wally Broecker of the Lamont-Doherty Earth Observatory at Columbia University (Broecker, W. S., 1991 and 1992) the Gulf Stream has not always flowed to the far north as it does today. When it has ceased to have a warm surface flow as far north, that sector of the world - Greenland, Iceland, northern Europe - has generally gotten colder. It has gotten enough colder that despite there being less water vapor present, what little snow that does fall will not tend to melt in the orbitally-controlled shorter summers and glaciers slowly will begin to build.

While changes in the patterns of major ocean currents may take many thousands to millions of years, this shift from a far-north-flowing Gulf Stream to a flow ending farther to the south appears, from the detailed studies of ice cores in Greenland and sea floor sediments in the North Atlantic, to have occurred in a much shorter time. This change may have been caused by large volumes of lighter, salt-free fresh water supplied by melting glaciers or ice sheets 18,000 years ago to make the saltier, heavier Gulf Stream sink below this lighter fresh water. See Figure 22 to observe the magnitude of the possible change in location of the northern extent of the Gulf Stream.

So what if it gets a little colder in Greenland and northern Europe? The problem is that there is currently no massive amount of artificial heat available to heat the fields and help grow crops in a much colder climate. This is an example of a large regional climate change. Don't forget that the interplay of many other factors (orbital, albedo, location of continents, greenhouse gas content, etc.) can act together to cause the onset of a world wide or regional cooling or warming. If the world's climate cools enough to affect the world's agricultural output significantly, the wars and chaos caused by trying to squeeze more than six billion people into a small band around the equator will make any previous world war insignificant in comparison because everyone on Earth will be fighting for what little productive land remains.

There are very short-term changes in some surface current patterns that modify Earth's regional climates. The shifts in the central Pacific from warmer El Niño surface currents to cooler La Niña surface currents are examples of changes that can and do occur every few years. Even these relatively small changes in ocean surface temperature cause changes in the climate from North and South America to Africa and either help provide moisture, or wreak havoc by causing drought, in certain areas of those regions.

Fig. 22

CREATION OF THE GULF STREAM THREE mya

Five mya, an open passageway existed between North and South America. Since this very warm water was not circulating northward, the north pole was very cold and dry, and glaciers grew very slowly because of the lack of moisture for snow production.

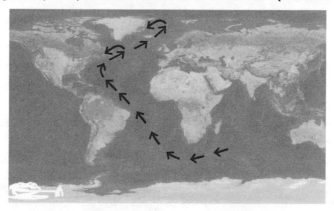

By three mya, tectonic forces caused land to be uplifted and close the passageway between North and South America, and the westward flow of warm water got diverted northward, where it delivered moisture necessary for large amounts of winter snow and rapid glacial expansion during all but the interglacial, warm cycles like the one we are living in today.

Plate Tectonics

The next several drivers are related to plate tectonics that cause a variety of horizontal and vertical continental movements that affect Earth's climate.

Location of the Continents

The location of the continents can make a big difference in the world's climate. As we have just seen, continent positions can block the paths or create new paths for the ocean currents. See Figure 4, page 4, for a generalized summary of the location of the continents and equatorial ocean currents at various times over the last two billion years.

Landmasses located near the poles can hold large volumes of glacial or sheet ice. During the most recent ice age, ice covered not only the North Pole but also vast areas of land over most of Canada and portions of what is now the northern United States and northern Europe. Many scientists have estimated that ice was approximately 1.8 miles thick just north of what is now New York City. The estimates are made by measuring the amount and rate of rebound (uplift) of Earth's surface from where it had been depressed by the weight of the overlying glacial ice.

Continents located directly over a pole can provide a cold, solid base upon which to build huge volumes of ice. Once land at the pole gets cold and covered with ice, it remains frozen until a major warming occurs. When water instead of land is present at a pole, the ocean circulation constantly delivers some heat to the base of any ice that has formed and the ice will not tend to build up nearly as thick as on land. Since more than 85 percent of the ice that floats in the ocean is beneath the ocean's surface, like the iceberg that sank the *Titanic*, the elevation of the ice will never be very high. This lack of elevation means that temperatures over a landless pole will be higher than where ice has built up to higher (colder) elevations on landmasses that overlie a pole. Temperature everywhere on Earth declines with increased altitude. As a current example, the continent of Antarctica at the South Pole holds more than seven times as much ice volume as that present at the open ocean of the North Pole, even when including the ice accumulated on the landmass of nearby Greenland.

When the landmasses covering part of the surface of Earth are broken into many scattered pieces, a relatively uninterrupted flow of ocean currents can occur. This allows heat to be distributed in latitudinal bands, with warm surface currents found at lower latitudes and cool surface currents found nearer the poles. When these perpetually cooler waters deliver little moisture to the overlying atmosphere, glacial and sheet ice grow very slowly. If the summers are relatively hot and melt the ice each year, no build-up will occur, unless the land's altitude results in permanent freezing.

Alternatively, when the landmasses coalesce into one or two supercontinents, or a large landmass is located directly over the equator, warm equatorial currents can be blocked and diverted poleward which takes more heat and moisture in that direction. This moisture is then available to feed the larger snowfalls and ice buildup.

The supercontinents also tend to harbor vast deserts in their interiors. These vast deserts that replaced earlier, plant-covered, small continents have a much higher reflectivity coefficient (albedo) than a plant-covered surface. While the desert surfaces are hot, I believe the net effect or more of the sun's energy being reflected back into space should result in a somewhat cooler average global temperature.

Continental Drift or Sea Floor Spreading

What causes the location of the continents to shift? It has been called continental drift, plate tectonics, or sea floor spreading. Below the continent, Earth's crust is approximately 20 miles thick, whereas below the sea floors, the crust averages only about 3.5 miles thick. The continental plates, several of which carry continental landmasses with them, "drift" or move around on a molten subsurface as a result of heat convection cells that rotate within this molten layer (forces caused by heat flow within Earth), as well as possible effects from Earth's rotation or gravitational influences by the sun, the moon or the other planets. Again, look at the Grand Summary chart, Figure 4, page 4, to see how the continental locations have changed over the eons.

The rate of the plates' movement is only an inch or two or three a year, but over millions of years, these distances can become thousands of miles. The separation of the Americas from Europe and Africa, which began approximately 175 million years ago, has been spreading at a relatively low rate compared with other plate movements but now has produced the Atlantic Ocean which is up to 4,500 miles wide between South Africa and lower South America.

Of course, the plates not only spread but also collide. Where these spreading rates and the accompanying collision rates are very high, Earth produces inordinately large earthquakes, tsunamis, volcanic activity, and mountain building. Even at a modest collision rate, the Himalayas are an example of exceptionally high mountain building that is being caused by the northward drift of India into the continental Asian plate. This collision began approximately 55 million years ago and continues today. The collision has caused the uplift of the 15,000-foot-high Tibetan plateau, which, along with the Himalayas, has a very strong effect on the moisture distribution over a large portion of that region. Another active area, one with a more rapid collision rate, lies under the ocean west of Indonesia where we frequently witness large earthquakes, tsunamis, and volcanic activity. You might remember reading of the huge volcanic explosion of Krakatoa in 1883, often referred to as "the shot heard around the world", and the very large Indonesian earthquake of 2005 that caused a tsunami that killed more than 200,000 people in the same general part of the world.

Where these truly huge masses of land (continental) or sea floor (oceanic) plates collide, one plate will override, and one will underride (get subducted beneath), the other (Figure 23). The overriding plate usually gets squeezed up into mountain ranges such as those along the North and South American west coasts, where those continental plates are overriding the oceanic crustal plates of the eastern Pacific.

Where one plate under-rides another, that crust is pushed down and heated to a molten state. Some volume is squeezed back up through the newly formed volcanoes as lava with its associated gases (again, see Figure 23). So, the whole plate-tectonic scene either can create new land or mountains, or it can destroy existing crust of Earth by recycling it back down into the molten layer of Earth with some of it being extruded back up through volcanic activity. Relating this phenomenon back to one of Earth's greenhouse gases, much of Earth's carbon dioxide is released into the atmosphere by this relatively constant volcanic activity, a factor we cannot control, and also by the degassing of the oceans when Earth's temperature rises. Today, some of the increase in CO_2 is occurring as a result of human-related emissions while the rise in Earth's temperature also is probably contributing to the increase of atmospheric CO_2.

Fig. 23

CONTINENTAL DRIFT, COLLISION AND SUBDUCTION

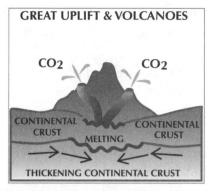

Continent-to-continent plate collision
(India and Asian continents)

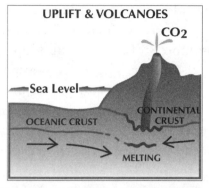

Oceanic-to-continental plate collision
(West coasts of N. and S. America)

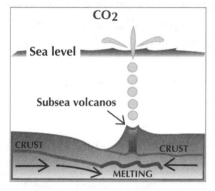

Oceanic-to -oceanic plate collision
(Philippine and Pacific plates)

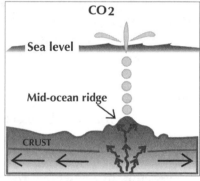

Mid-ocean spreading center
(mid-Atlantic ridge)

H. Leighton Steward, 2007

At the mid-ocean spreading centers, when spreading rates are high, the rising molten rock causes the sea floor to swell along these multi-thousand-mile long ridges. This significant elevation of the sea floor can displace water onto the continents and alter (lower) Earth's reflectivity factor or albedo. Alternatively, with low spreading rates and a much lesser swell at the spreading centers, sea level will tend to be lower and Earth's albedo higher as a result of more land surface being exposed. Also, CO_2 coming from the spreading centers will be more modest and result in a lower greenhouse effect.

What does all of this have to do with climate? A lot! We have seen that continental locations and size can affect ocean currents, location and extent of deserts, and plant life. Also, continents provide bases upon which to build huge volumes of ice, including the effects of increased elevation and the resulting cooling to promote still more ice buildup.

Elevation also affects atmospheric circulation and rain or mountain snow and glaciation. Most of us know that when a wind blowing from west to east encounters a mountain range, the upper west flank and the top of the range will get more moisture in the form of rain or snow because of the cooling effect on warm air rising. Since cold air cannot hold as much moisture as warm air, the moisture is released. On the east or downward flank of the mountains, most of the moisture already has been precipitated, plus the air is getting warmer again as it descends and can retain what moisture is left and not be precipitated as rain or snow. As long as the mountain ranges exist, the leeward side of the ranges will be drier.

The areas that are elevated or built into mountains can cause various rates of erosion and chemical weathering that may have climate effects that will be discussed later in this chapter.

Ultimately, the active plate movements and continental collisions that build or replace eroded land are the only reasons we have land upon which to live in this paradisial climate we are currently experiencing. Otherwise, the mountains would eventually all erode away, and we would have a shallow layer of water covering the entire Earth! Never thought you would look favorably at land building by earthquakes and volcanoes did you?

Elevation of Land Masses

The height to which land is elevated - high mountains or high plateaus - not only affects atmospheric circulation and moisture distribution but also has an effect on physical and chemical weathering rates. As new rock or sediment surfaces are exposed to the atmosphere, chemical weathering causes a reaction with the siliceous rocks (quartz-bearing sandstones or granitic rocks) and, in the process, removes large volumes of CO_2 from the atmosphere and that may help cool the atmosphere somewhat. Since spreading rates of the sea floor ridges and the accompanying subduction zone trenches, tend to average out over time, the very long-term effect on CO_2 should be fairly constant. However, sometimes instead of the oceanic plate colliding with each other, two or more continental plates will collide and cause an inordinate rise in elevation of a large area, such as the Himalayas, and greatly increase the rate of removal of CO_2 from the atmosphere.

What causes this increase in chemical weathering? Steep slopes collapse and break up easily, earthquakes are quite prevalent in the mountain-building phases that last for tens of millions of years, water gets between the cracks or fractures in the rock and freezes and mechanically breaks up the rock, and glaciation grinds up large volumes of rock into much smaller fragment and exposes many more square centimeters of fresh surface to be chemically altered.

About 55 million years ago, the plate bearing the continent of India began to collide with the continental Asian plate. The collision is still occurring. The collision continues to build an unusually high plateau that is being actively eroded, the Tibetan plateau. The collision also has built the tallest perpetually cold mountains of Earth, the Himalayas, which are still slowly growing in elevation. Look at Figure 24 and observe how much the temperature and average CO_2 level have declined since this uplifting began. Other ideas for this long-term decline in temperature and CO_2 include the carbon being stored in new "limestone" rocks and the lower sea floor spreading rate of the Tertiary period having pumped less CO_2 into the atmosphere and also causing sea level to be lower. Lower sea level results in more land exposure and a higher reflectivity of the sun's energy back into space. Or, was the temperature decline a cosmic ray effect that will be discussed near the end of this chapter?

Fig. 24

TERTIARY TEMPERATURE AND CO2 DECLINE

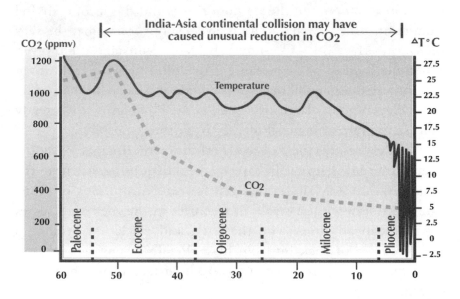

Temperature and CO2 levels have both declined over the last 60 million years.

After Bluemle et. al., 2001 and Berner, 2004

No major collision had occurred for the previous 185 million years, and Earth's temperature and CO_2 content had become very high in this quiet interval. Spreading rates and the resulting sea levels were also higher 100 million years ago. Most scientists believe that little permanent ice existed during this mid-Cretaceous warmth. Farther back, between 325 and 240 million years ago, many continent-to-continent collisions occurred which allowed elevation of landmasses and much more chemical weathering and long term CO_2 removal. Also, with the supercontinent named Pangaea blocking equatorial current circulation which helped with additional precipitation of snow and a higher albedo, Earth cooled and evidence shows that ice covered the poles and extended to latitudes below 40 degrees. This make an argument for having some long-term effect by CO_2 and therefore some possible impact on global climate unless the reduction in CO_2 was simply caused by the substantial drop in temperature during the ice age of that time, just as CO_2 levels have been known to follow the temperature decline of the last several glacial cycles. The latter seems more plausible.

Chemical and Mechanical Weathering

Some paleoclimatologists believe that mechanical weathering rates, which affect chemical weathering, can have a long-term influence on Earth's climate. The principal reason is that mechanical weathering, such as that caused by elevation of land through plate collision, exposes a lot of new rock debris that can be acted upon by chemical weathering and lower the carbon dioxide content of the atmosphere. The degree of this effect is still being debated within the scientific community.

In addition to simply exposing more rock surface to the atmosphere and the associated chemical weathering, mechanical weathering from wind, water, ice, and earthquakes can physically move more rock and soil to the sea that can get turned into carbonate reefs (turning more carbon into rocks) and carbonate seashells like foraminifera, or into siliceous, quartz type shells such as diatoms. The microscopic diatoms are useful not just as a food source for larger fauna in the oceans, but as the diatomaceous Earth that coats the water filters that keep swimming pools crystal clear.

The general cooling of Earth's climate over the last 55 million years has caused additional CO_2 to be absorbed by the cooler oceans, which is another reason for the CO_2 decline. Regarding the oceans, in our current warm period of paradise, the recent warmer atmosphere and resulting increase in ocean surface temperature has caused more CO_2 to be released from the sea and into the atmosphere much like a warm bottle of soda release more CO_2 than a cold bottle of soda.

However, as Earth continues to warm and the ice sheets and mountain glaciers melt and expose long-buried bedrock, the process of chemical weathering on these surfaces can resume and help slowly remove CO_2 from the atmosphere. Less CO_2 results in slower plant growth which means less of the greenhouse gas methane being created and less plant cover allows for a more reflective land surface and a cooler Earth. These are examples of other feedbacks that help prevent a runaway greenhouse warming. Earth has experienced runaway glacial periods ("Snowball Earth" periods) but never runaway hothouse periods, even at 7,000 *ppm* CO_2.

Speaking of CO_2 causing runaway global warming, a close friend's daughter came home and said her teacher told her that if humans keep putting CO_2 in the atmosphere at the current rate the Earth will be destroyed in 75 years! Keep that in mind as you continue reading.

Extraterrestrial Impacts

Look at the surface of the moon to get some feel for the number of times Earth has been hit by objects from space. With the Earth having a much larger cross section in space and also having a much larger gravitational attraction on approaching space objects, Earth has been hit many more times than the moon. We can catalogue many impact craters on the surface of the land or the sea floor but they are just a minuscule number compared with the total that has penetrated our atmosphere (Figure 25). Why don't we see all their scars? Erosion gradually wears away their rims and then the craters themselves or fills their centers and then covers everything with a new layer of soil or rock. Also erasing the scars is the effect of plate tectonics that continually destroys (subducts) old oceanic crust. With 70 percent of Earth's surface covered with water, most of these tsunami-creating impacts occurred in the oceans. Meanwhile, new crater-free oceanic crust is being created at the mid-ocean spreading centers by the out-pouring of new molten rock. For an excellent data source of where nearly 200 of the impact craters are located on Earth's surface, refer to the University of New Brunswick (2008) source, referenced in the Bibliography.

Most objects that enter Earth's atmosphere are traveling between 40,000 and 60,000 miles per hour (11 to 16 miles per second!). Most rifle bullets only travel at 2,000 to 3,000 <u>feet</u> per second. If just an auto-sized boulder struck your neighborhood at 60,000 miles per hour it would wipe out not only you and your neighborhood but likely your entire city and county.

The most recent space rock of super size (about six miles across) struck Earth in just such a manner 65 million years ago and apparently helped kill off the dinosaurs. The crater it left beneath the Yucatan Peninsula in Mexico has been measured at approximately 110 miles across. It has been speculated that the wave it caused hundreds of miles away, still had to be more than 1,000 feet high as it crossed above Florida to account for the size of the boulders the water carried into the western Atlantic Ocean.

Fig. 25

EXTRATERRESTRIAL IMPACT CRATERS

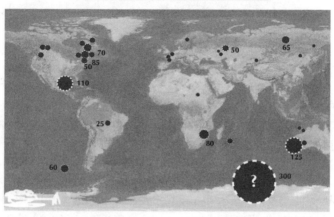

Known impact craters more than 15 miles in diameter with mileage shown for all more than 25 miles across. More than four times this number of craters have been identified with craters less than 15 miles in diameter. Scale is relative only crater to crater. Why didn't the huge, instant input of CO_2 into the atmosphere caused by the vaporization of carbonate rocks trigger runaway warming if, as some claim, CO_2 remains in the atmosphere for up to 200 years?

Personal communication with D. Abbott, 2008
and R. Von Frese, 2006

The impact, resulting heat wave, wild-fires, and mega-tsunamis undoubtedly killed millions of dinosaurs but the major extinction event was probably the sudden climate change caused by the dust blocking the incoming energy from the sun, causing a "nuclear" winter that persisted for several years. Plants died from cold and/or lack of sunshine. Also, there would have been a large amount of acid rain triggered by the impact. Volcanic activity was high at this time, probably caused or aided by, the cataclysmic impact. All of these deadly effects caused by the impact contributed to the extinction of more than 60 percent of the species on Earth.

While many would say this impact-driven climate change was temporary, its destruction of much of the world's continental fauna and flora, as well as a lot of sea life that previously interacted with the carbon content of the atmosphere, must have had some lingering effect on climate. Since CO_2 levels have fallen gradually since the impact, the fauna and flora obviously rebounded within a reasonably short geological period of time and the flora, through photosynthesis, helped remove some of the CO_2 that was released when the 110-mile hole was blasted into the thick sequence of carbonate rock ($CaCO_3$).

Even larger, older impact sites have been found on Earth. All of the large ones, except for those before life appeared on Earth, seem to be coincident with periods of climate or atmospheric changes and the extinction of many species. While the long-term climate change may not seem to be as significant as that brought about by some of the key long-term climate drivers, the short to intermediate effects can be very dramatic. Given some positive feedback from other drivers, impacts may have helped move the world into a new climate state.

New impact sites are continually being discovered. Researchers from Ohio State University, including Professor Ralph Von Frese (Von Frese, Ralph, 2006) using high-resolution gravity data, have found what appears to meet some of the criteria for an impact crater. Located about a mile beneath the sea floor and under an ice sheet near the coastline of eastern Antarctica (Wilkes Land), the anomaly measures approximately 300 miles wide. If it does prove to be an impact crater, the object from space that created it is calculated by Professor Van Frese to have been about 30 miles wide! The location of this anomaly is shown with a question mark in Figure 25. Should the estimated date of about 250 million years ago be correct, it is probably the smoking gun of the event that ended the Permian period and resulted in the greatest extinction of life ever recorded on Earth. It would have had an impact on global climate, both immediately and, through eliminating plant and sea life, for thousands of years to follow.

Another possible impact crater, about 125 miles wide, that occurred about 250 million years ago, has been located off the continent of Australia. Did a huge comet or asteroid break up and hit Earth with a double whammy? More research probably will tell us.

A much smaller and more recent impact crater "only" 18 miles wide has been found in the Indian Ocean about 900 miles southeast of Madagascar in 12,500 feet of water. Huge, unusual, wedge-shaped deposits call chevrons are found onshore that "point" toward the reported impact site. These instantly deposited wedges of sediment stand 600 feet high, are found three miles inland, and contain fossils of deep-water origin that are fused with metals typical of impact sites. The tsunami is estimated to have been more than 600 feet high to account for the elevation of the chevrons. Compare this with the 2005 earthquake caused tsunami in Indonesia, which was less than 100 feet high.

Dr. Dallas Abbott, a research scientist at Lamont-Doherty Earth Observatory at Columbia University and a leader in the research on this recent crater, reported in the *New York Times* on November 14, 2006, that the time of impact was thought to be about 4,800 years ago (Abbott, Dallas, 2006). While Earth did experience a drop in temperatures that lasted several hundred years about 4,500 years ago, there is no current evidence that the two events are related, but time will tell. More evidence of its effect on Earth surely will be found. To put the impact in historical perspective, this was just before the great pyramids of Egypt were built. That something from space struck Earth that recently and created a hole 18 miles wide beneath 12,500 feet of water further legitimizes researchers' attempts to figure out how we might deal with the next such object headed Earth's way.

Dr. Abbott is also a champion of the idea that the huge extraterrestrial impacts likely caused periods of enormous volcanic eruptions, a hypothesis that I also prefer unless new data prove it incorrect. Dr. Abbott says the statistical analysis by the group at Lamont has determined, with 97 percent confidence, that nine of ten of the largest volcanic fields that have been identified in Earth's history have been coincident with a large extraterrestrial impact. See Figure 25 again, page 51, for the location of identified impact craters more than 15 miles in diameter. There have been more than 125 impacts identified on Earth with craters less than 15 miles in width.

Volcanism

Volcanic eruptions also can affect Earth's climate and the extent of the effect is generally proportional to either the magnitude of the eruption, the duration of the eruption, or the type of the eruption (explosive versus simply flowing). Examples range from the relatively gently flowing Mauna Loa volcano in Hawaii, to a somewhat explosive Mount Saint Helens in the northwest United States, to Krakatoa in Indonesia, which in 1883 blew the entire top off a mountain and caused Earth to go without summers for a couple of years. Volcanically ejected glass from Krakatoa is present in the growth bands of star corals located in the Florida Keys. (Shinn, E.A., 2009, personal communication) Then there was the Toba volcanic explosion, about 74,000 years ago in

Indonesia, which blew 200 cubic miles of debris into the air and almost wiped out our ancestors. Some geneticists have speculated that only about 10,000 people survived the Toba event and nearly all of them were located in one region in Africa. That certainly would make us all not-too-distant cousins who have chosen to have serious family fights rather than just getting along with each other. Evidence elsewhere indicates other isolated pockets of humans survived.

Two hundred and fifty million years ago, a vast volcanic field in Siberia, the Siberian Traps, covered an area the size of Alaska. This volcanism, which was approximately coincident with another suspected giant impact(s), changed the makeup of the atmosphere and probably caused or helped cause the extinction of most species on Earth. One more example of volcanism would be the mega-eruption at what is now Yellowstone Park which occurred about 630,000 years ago. That super volcano exploded over a relatively short period of time, dumped ash in the Pacific Ocean and the Gulf of Mexico, and spread a layer of ash a foot thick 750 miles away in Iowa (Smith R. and Siegel, L. J., 2000).

These larger blasts or long lasting volcanic eruptions cause the climate to change in interesting ways. The initial blast of dust, debris, and small sulphate particles immediately cools Earth by blocking (reflecting) the sun's rays from reaching Earth's surface. A few years later, however, other climate drivers take control again and help cause Earth to continue warming or cooling. Long-term vast but nonviolent eruptions that continuously put more CO_2 and reflective sulphate particles into the atmosphere affect the climate in opposite directions and the net effect is not yet well known. What exactly happened, and when and why, are in the Earth's "library" just waiting to be discovered by some paleoclimate researcher.

Albedo

Albedo is a measured percent of the reflectivity of Earth's surface at any given time or sun angle. Clean snow or ice or pure white clouds have a very high albedo of around 90 percent for direct overhead sunlight, while the lowest albedo of around five percent that absorbs most of the sun's energy occurs in dark forests, dark soils, and sunlight striking the oceans or large bodies of water from directly overhead (Figure 26).

As the snow gets dirty, the clouds get darker, and light colored soil gets covered with grass, plants, or water, the albedo will drop. While it is easy to imagine snow or ice having a high albedo, direct overhead sunlight striking the desert or a white sandy beach also can have a high albedo (as high as 50 percent).

Albedo is a very important component because it determines how much of the sun's energy is absorbed at Earth's surface and is very important in affecting Earth's climate. Cloud cover, the amount of water versus land, and the amount of plant cover versus desert surface that the sunshine strikes from a high angle are probably as significant to the total Earth albedo as are the volumes of very high-latitude ice and snow cover. An exception to this ice and snow cover effect would have occurred when the glacial periods experienced ice cover all the way down to the mid-latitudes. To visualize the albedo-changing effect of a past high-sea-level stand that inundated many continents where low-albedo water covered previously higher-albedo land, see Figure 27, page 56.

Fig. 26

ALBEDO

EARTH'S LOW-ALBEDO SURFACES

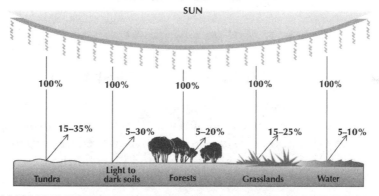

EARTH'S HIGH-ALBEDO SURFACES

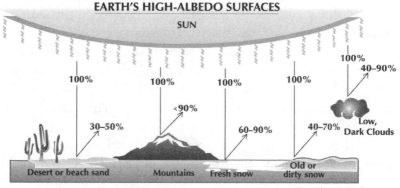

H. Leighton Steward

Fig. 27

SEA LEVEL 100 MILLION YEARS AGO

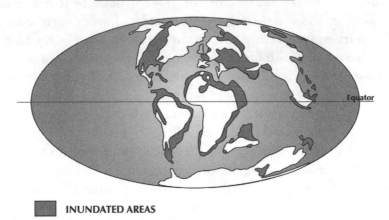

INUNDATED AREAS

High sea level occurred when there was no permanent glacial ice and when seafloor spreading rates were higher than today. Earth's albedo dropped as water replaced the land surface, causing more warming.

Adapted from D. Merritts et al., 1998

Flora and Fauna (Plant and Animal Life)

Grass, shrubs, and trees absorb a lot of sunlight which results in a low albedo and warmer land surface. These same plants take the greenhouse gas, carbon dioxide, from the air and release oxygen. This then tends to help cool the atmosphere and keep Earth at a relatively moderate temperature level. As stated earlier, an abundance of plants and trees that live and die in swampy areas generates a lot of methane which is also a greenhouse gas that will tend to capture some of the infrared heat radiated from Earth's surface.

I have heard that people say their plants grow better if they talk to them every day. While this old saying is thought by many to be an "old wives' tale", it underscores the benefit of CO_2 to plants because we release CO_2 with every breath we exhale. Animal life, including people, takes in oxygen and gives back CO_2.

To digress a moment, the Supreme Court of the United States recently ruled that the Environmental Protection Agency (EPA) has the authority to rule that CO_2 is a "pollutant" and can be regulated.

One could ask why the same rule does not apply for oxygen which can be toxic to pilots who breathe 100 percent oxygen for extended periods of time. People breathe in oxygen and plants breathe out CO_2. If the people who believe "CO_2 is bad" are successful in causing Earth's people to try to reduce severely the amount of CO_2 in the atmosphere, we will not have to worry about robust plants in our houses and agriculture in our fields, nor an overpopulated Earth, because there will not be enough to eat! With no scientists on the Supreme Court, it is little wonder that the judges come up with nonsensical views based on "evidence" provided by a very politically motivated Intergovernmental Panel on Climate Change (IPCC). The country in the world that stands to lose the most in economic strength by following a set of rules voted in by foreign officials is the United States. Do you doubt that they would mind seeing the United States reduced in world power and influence? The IPCC, at the highest levels, is made up predominantly of social scientists and political appointees. Many, many well known scientists have become disgusted with IPCC ignoring scientific input that conflicts with IPCC's Summary for Policy Makers.

Sea life is also important in Earth's carbon cycle because many of the forms of plankton and algae absorb carbon from the waters to make their calcareous shells. As the life forms die, the shells sink to the bottom and effectively sequester (store) the carbon there for long periods of time. This can lower atmospheric CO_2 and tend to have a modest cooling effect.

Meanwhile, volcanic activity associated with subduction zones or the sea floor spreading centers does release new CO_2 into the atmosphere. Over geologic time, more CO_2 has been sequestered in rocks than has been put back into the atmosphere. The "proof" of this lies in the fact that CO_2 levels four or five hundred million years ago were several thousand parts per million versus only 380 parts per million today.

Atmospheric Circulation

Just as the ocean currents distribute heat and some overlying moisture around the surface of Earth, the winds also affect where heat and moisture are distributed. Winds blowing across the oceans above the El Niño and La Niña currents have regional if not global effects on the climate for periods of several years. Winds drive surface currents that can create upwelling of deeper waters that provide the nutrient-rich environments for the growth of smaller sea life. This microscopic life removes CO_2 from the atmosphere by incorporating the carbon in their shells. These nutrient-rich zones also support an abundance of fish that feed on these smaller animals.

Winds blowing across continents play a very important role in distributing moisture to areas far removed from the coastlines. Since atmospheric moisture content affects cloud formation and precipitation, winds can affect temperature and vegetation patterns. While wind patterns are usually a more important factor in regional climates, these regional climates all add up to affect global climate. Wind enhanced rainfall also causes more chemical weathering, which, along with vegetation, can remove CO_2 from the atmosphere. For a very simplified illustration of Earth's major atmospheric circulation patterns, see Figure 28.

Fig. 28

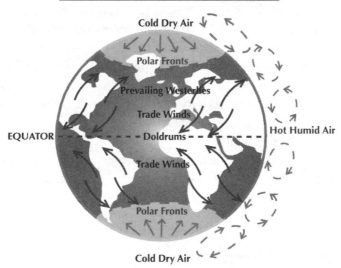

ATMOSPHERIC CIRCULATION

A very simplified map of Earth's atmospheric circulation.
The amount of wetness or dryness taken up by these cells depends on whether the cells are over the oceans, deserts, etc., when the air is rising.
This can affect where precipitation occurs and also the moisture content of the descending air.

Earth's Rotation

The rotation of Earth is not often listed as a climate driver but I think it deserves mentioning. An Earth with no spin would mean a half-fried, half-frozen Earth. The half that received sunlight gradually would change at the rate of $1/365^{th}$ of the surface each day as Earth made its annual orbit of the sun but the result would be essentially the same; one half would be temporarily fried, while the other side would be temporarily frozen.

The spin of Earth affects the atmospheric circulation patterns as well as the oceanic circulation patterns, which we know are both distributors of the heat that affects Earth's climate.

The rotation is *"constant"* (in quotes only because Earth's rotation is gradually slowing down). After the collision that gave birth to the moon, Earth's rotation was faster. A day on Earth lasted only about six hours. Don't try to use this bit of trivia in calculating tomorrow's climate change; it isn't worth it! The 16 hours of slowing has taken 4.5 billion years in which to occur.

Cosmic Rays

I first heard of cosmic rays when I was very young. Comic-strip and Saturday matinee characters used cosmic ray guns as they traversed the universe. While all this was very intriguing, I had no idea that cosmic rays really existed. Well, cosmic rays really do exist and it has been proposed that they are important in determining when many of Earth's global climate changes occur. Rather than being shot from fictitious cosmic ray guns, these truly atomic bullets originate and get sprayed through space when suns, much more massive than our own, ultimately collapse and explode with unimaginable force (a supernova). With trillions of such suns in the universe, supernova explosions occur very frequently. Most of the explosions happen far, far away and have no discernible effect on Earth but some blow up "nearby" and shower our solar system with cosmic rays.

What we are interested in is what effect, if any, these cosmic rays have on Earth's climate. Dr. Henrik Svensmark, (Svensmark,

H. and Calder, N., 2007), director of the Center for Sun-Climate Research at the Danish National Space Center, and other scientists such as Jan Veizer of the University of Ottawa and Nir Shaviv of the Hebrew University of Jerusalem (Shaviv, N. J. and Veizer, J., 2003), believe the answer is that cosmic rays do have a significant effect on the climate by controlling the amount of low-level cloud cover on Earth. Svensmark has backed up his beliefs with some fairly crude but nevertheless encouraging experiments that seem to support the hypothesis that cosmic rays that survive the trip through the sun's and Earth's magnetic shielding and Earth's upper atmosphere, cause ionizations as the rays, traveling near the speed of light, collide with molecules in the lower atmosphere. These newly created, very small particles then provide the nuclei on which water droplets can form, which then can create low-level clouds. The low-level cumulus clouds have highly reflective tops than can and do reflect much of the sun's energy back into space and, in doing so, help cool Earth. The greater the volume of comic rays that reach Earth's lower atmosphere, the more collisions would occur and create more small particles that seed cloud formation and the more clouds would form to provide this cooling. Conversely, fewer cosmic rays reaching Earth's lower atmosphere would mean fewer clouds and more sunlight reaching Earth's surface, which would then cause Earth to warm.

Most cosmic rays that enter the solar system do not make it to Earth's surface. They are repelled predominantly by the magnetic field of the sun and its solar wind. They are also repelled to a lesser extent by Earth's own magnetic field. Still other cosmic rays are eliminated as they collide with molecules in Earth's upper atmosphere where the clouds that form there do not have the same cooling effect because the high clouds are mainly made of ice crystals that tend to hold heat in the super-cold environment of the upper atmosphere. To see how the sun's activity level, as measured by the number of sunspots present, helps determine how many cosmic rays reach Earth, see Figure 29. Also note how closely the cosmic ray density and percentage of low cloud cover correlate.

Fig. 29

A. COSMIC RAYS AND SUNSPOT NUMBER COMPARISON

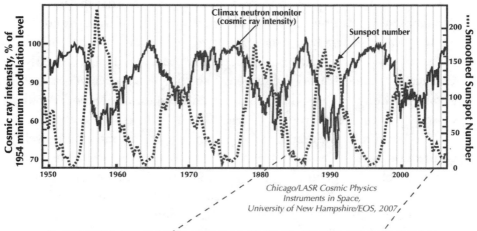

Chicago/LASR Cosmic Physics
Instruments in Space,
University of New Hampshire/EOS, 2007

B. COSMIC RAYS AND GLOBAL CLOUD COVER COMPARISON

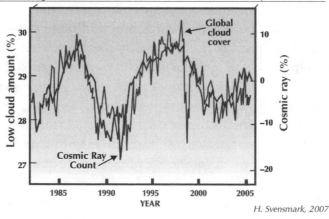

H. Svensmark, 2007

Fortunately, cosmic rays that do enter Earth's atmosphere leave both modern and ancient footprints or indicators to tell us of their relative abundance. Here the excitement builds, since the old footprints tell us, so far, that the volume of cosmic rays reaching Earth was unusually high during times when Earth was experiencing major glaciations. When more cosmic rays are present, the production rate of atoms of the stable isotopes beryllium-10 and carbon-14 rise and allows researcher to measure the number of these atoms that are in Earth' library. An anomaly in this otherwise consistent picture has been that the Jurassic and earliest Cretaceous periods, that were originally thought to be ice-free, were generally a time of active cosmic ray bombardment. Recently, however, evidence of glaciation in Australia during the early Cretaceous Period has been discovered, so the gross correlation is still possible although weaker than during other ice ages. See Hypothesized Cosmic Ray Impact Summary on page 63.

One reason for the excitement regarding the cosmic ray-cloud cover connection is that no one driver identified to date has stood the test of being able to explain all the major changes in Earth' paleoclimate. Although no one expects to find a single do-it-all driver, everyone would like to see another driver that might explain some of the major climate changes. Will cosmic rays withstand more detailed scientific scrutiny? Will evolution of some of the life on Earth be tied to cosmic ray influx as Dr. Svensmark suggests? Only time and more research will tell.

A major research facility to test the hypothesis that cosmic rays can help cause the formation of clouds is located at Europe's world-class particle physics laboratory called CERN in Geneva, Switzerland. Unfortunately, with start-up problems, it may be two or three more years until significant results can be expected from this facility. In the meantime, more research will be undertaken to try to document the times on Earth when cosmic ray volumes were abundant or sparse and compare those abundances with the temperatures at those times.

Let's assume that cosmic rays are shown to have a measurable effect on global climate. Can we hope to predict either when more comic rays will appear from space or when the solar shielding will be strong or weak enough to affect how many rays survive to Earth's lower atmosphere where they might create more clouds? The answer to the abundance from space is "probably not" for a specific time because how can we predict when a comic-ray generating supernova will occur "close by"? How can we see the rays coming if they are traveling near the speed of light? We are able to predict when our solar system which is part of the Milky Way galaxy, will orbit through one of the spiral arms of the Milky Way, where these massive suns (stars) are more abundant and where more near-by supernovas would be expected to occur. We are near the outer limits of one of these more star-dense spiral arms, the Orion Arm, and we are in an extended period of glaciations except for our temporary, warm, interglacial paradise. Relative to predicting the sun's future magnetic-shielding activity, we are just beginning to forecast changes in magnetic activity and more observations and research may give us the ability to make some meaningful predictions.

What if we find we cannot predict the periods of comic ray intensity variations or of the sun's major magnetic cycles? We should then spend our research time and dollars in determining what we should do to best react to any such changes that might occur. In my opinion, we should be spreading our research efforts and dollars studying all of the causes and effects of all the climate drivers, not just focusing on the as-yet-to-be-proven climate-changing effect of carbon dioxide.

Is it likely that one "silver bullet" will be found that controls all observed climate changes? I don't think so, because all the drivers make some contribution, and any one, or any group of drivers, can have a modifying effect on the climate at any particular time and in either a positive or negative direction. We probably have progressed only part way down one aisle of the very large Earth library of paleoclimate history.

Hypothesized Cosmic Ray Impact Summary:

A.   Cosmic rays from space approach the solar system.

B.   Solar magnetic shielding varies in strength.

  1. Low magnetic field strength (few sunspots and weak solar wind) allows many cosmic rays to enter Earth's atmosphere.

    a. Cosmic rays collide with particles in the atmosphere, ionize them, and provide nuclei for formation of water droplets and clouds.

    b. Clouds formed in the lower atmosphere have highly reflective, high albedo, tops that reflect a high percentage of sunlight (heat) back into space.

    c. More clouds mean less solar heat reaches Earth and Earth becomes cooler.

2. High magnetic field strength (many sunspots and strong solar wind) deflects most cosmic rays away from Earth and fewer low-level clouds form and more of the Sun's heat is absorbed at Earth's surface and Earth warms.

Driver Commentary

What do you think is causing the climate to change? Is it still just carbon dioxide? No. Is carbon dioxide important as a temporary retainer or radiated heat? Yes, particularly at low concentrations (Figure 16, page 28). Are there other factors we should take into account and study besides carbon dioxide that we might be able to influence in some favorable way and at some reasonable, nondestructive pace? Yes. I will mention some possibilities in later chapters.

Chapter Four

CLIMATE INDICATORS

This chapter addresses your reasonable question, "How in the heck do you scientists know what happened, say, 430 million years ago?" If you really want to understand how we know these things, read on. If you are very trusting of the great number of scientists who have tested these methods for their accuracy and interactions and want to get back to more discussion on climate change, just browse on.

Indicators paleoclimatologists use to determine ancient climates and when they occurred can be found in Figure 30, page 66. A paper clip for easy referral to that summary page might be useful.

To be able to use in a meaningful way what a particular climate indicator or suite of indicators is telling us about an earlier climate, we must know the time at which the particular climate was occurring. The most common methods used to determine dates rely on the decay rates of certain naturally occurring radioactive elements that were formed at the same time as the medium in which the minerals are found. Each type of radioactive element has its own specific half-life, the time it takes for one half of the original number of radioactive atoms of an element in a sample to decay. Knowing this half-life and measuring the number of radioactive atoms of an element remaining in a sample allow a determination of how long it has been since the element in the sample first formed. With this knowledge, we can date the particular rock or sedimentary layer from which we derive our atmospheric and climate information. Radioactive elements commonly used to determine the dates of geologic materials are shown in Figure 31, page 67.

Stable Isotopes

This is a biggie. The best current method we have to pry useful information from the many indicators is stable isotopic analysis. It is providing answers to what we most want to know, which are what the temperatures were at the time, how much CO_2 and oxygen were present, and the relative amount of measurable trace gases or cosmic rays that impacted Earth's atmosphere. Having said that I have attempted to write a book about global climate change and the factors that affect it for the layman on the street, it is time for a short summary on a highly technical area regarding stable isotopes.

Fig. 30

CLIMATE INDICATORS

INDICATOR	WHAT WE LEARN FROM IT	ITS LIMITATIONS
Thermometers	Exact temperature at that site.	Available less than 200 years worldwide.
Ice cores	Temperature, CO_2 , and methane. Dust and sea-salt volumes for wind velocity.	Covers a maximum of 850,000 years at the south pole.
Tree rings	Hot vs. cold and moist vs. dry periods and specific age dates.	Only applicable up to a few thousand years.
Lake and bog cores	Plant and pollen, sediment variation, and dust (wind velocity).	Rarely find good samples beyond 25,000 years.
Ocean cores	Plant, pollen, shell type, dust abundance, latitudes of glacial sediment, volcanic activity, and extraterrestrial impacts.	Ocean floor erased by continental drift (subduction) beyond 175 mya.
Glacial footprints	Distribution and timing of glacial or sheet-ice on Earth.	Availability of glacial deposits or striation marks.
Vegetation and pollen	Distribution of tropical to arctic trees, shrubs, and grasses.	Only availability of samples to examine.
Paleoshoreline positions	Inundation of continents, seafloor spreading rates and sea level.	Availability of widespread samples to examine.
Coral type and distribution	Temperature implications of where corals grew and paleoshoreline locations.	Only availability of widespread samples to examine.
Paleosoils (Paleosols)	Vegetation type, moist or dry soil type, atmosphere and water chemistry.	Availability of widespread samples to examine.
Fossils	Implications of warm or cold fossil types and age environment.	Only availability of widespread samples to examine.
Rock, sediments & stalagmites	Environment of deposition, e.g., desert evaporites or tropical rain forests.	Only availability of widespread samples to examine.
Stable isotopes	Paleotemperature and atmospheric and fluid composition from many of the above.	Accuracy sometimes affected by changes postdating mineral formation.
Archeology	Climate implications of where, when, and how our ancestors lived.	Only robust sites to examine.
Written records	Eye-witness climate observations.	Writing only available for last 5,000 years.
Atmospheric sampling	Exact amount of gases and particulate matter in the atmosphere.	Only pertains to the last couple of centuries.

H. Leighton Steward, 2007

Fig. 31

NATURALLY OCCURRING ELEMENTS
COMMONLY USED IN AGE DATING

PARENT ISOTOPE	DAUGHTER ISOTOPE	USEFUL BACK TO	USED FOR DATING
Carbon–14	Nitrogen–14	< 50,000 years	Carbon bearing materials
Thorium–230	Radon–226	<400,000 years	Corals
Potassium–40	Argon–40	>100,000 years	Basalts
Rubidium–87	Strontium–87	100 myrs	Granites
Uranium–235	Lead–207	>100 myrs	Various Rocks
Uranium–238	Lead–206	>100 myrs	Various Rocks

The known radioactive decay rates of the half lives of the isotopes shown above allow calculations to be made using the remaining radioativity to determine the age of the material being examined. Without this ability, our estimates of the age of old rocks, etc., would be based solely on fossil evidence (if any) and result in only a gross approximation of the relative ages.

All of the atoms of a particular element have the same number of protons in their nucleus. Atoms of elements can, however, have a different number of neutrons that then yields a different mass or mass number. The same basic element but with a different mass number is called an isotope of that element. For example, the most common form or isotope of oxygen is ^{16}O, which has a mass number of 16. Another less common form or isotope of oxygen is ^{18}O, which has a mass number of 18. These isotopes come from the same element but are found in different proportions to one another depending on what environmental conditions exist at the time of their precipitation into the solid or fluid being formed.

^{18}O is heavier than ^{16}O. Both are present in water vapor, rain droplets, ice, or snow. The isotopes form the same chemical compounds but because they have slightly different physical properties, such as their melting and boiling points, the heavier isotope, ^{18}O, is favored in the condensation phase and, as you might expect, ^{18}O is precipitated out of the atmosphere to a greater degree than the lighter ^{16}O. A change in temperature or an effect from the distance traveled after the isotopes were formed can cause the ratio of ^{18}O to ^{16}O to change. As the temperatures continue to drop, more of the ^{18}O drops out and further reduces the ratio of ^{18}O to ^{16}O. Temperature has the dominant effect, but there is the effect from the distance traveled in the atmosphere. Thus, using other knowledge of temperature and conditions at that time, you sometimes can infer approximately how far the moisture may have traveled before it was precipitated in the form of rain, ice or snow.

Still other elements have isotopes that can be used by studying various ratios of their isotopes and comparing the effects of current, known factors to the ratios that exist in ancient samples of fluids or solids. This is a marvelous tool that has been used for several decades but is just in the dawn of its application for paleotemperatures and atmospheric makeup relative to the vast amount of material in Earth's library. Isotopic analysis provides information on a large array of minerals and the temperatures that existed at the time of their formation.

Ice Cores

Detailed ice-core analysis covering the last 400,000 years has provided an excellent record of paleotemperature for that period. Less reliable but fair climate resolution can be obtained from ice cores that date back to around 650,000 years ago, and gross estimates are possible from studying an ice core dating to 850,000 years ago. As the annual layers of snow and ice get more compacted and eventually folded and squeezed horizontally, the timing of exactly when a layer in the ice was laid down is less accurate. In addition to stable-isotope measurements that provide the paleotemperature values, the cores can provide atmospheric content for carbon dioxide to at least 200 meters of burial and also the amount of oxygen, methane, nitrous oxide, ozone, sulphur dioxide, and other trace gases or minerals. The volumes and origin of the dust also can be measured to help infer how windy the atmosphere was and to get a good measure of the general moisture content of the atmosphere. The snow and ice buildups from Antarctica and Greenland have preserved a climate record that can be analyzed in ways that would have been only a wild dream 50 years ago.

The annual snow layers are extremely well preserved for the last 10,000 years and can be individually counted and analyzed. As the snow gets further buried and the snow compacts and becomes ice, the ability to isolate annual layers, even in undisturbed ice layers, diminishes. Good estimates of total layers within an interval can be made, however, and generalized interpretations can be made for periods that are accurate to form a few hundred to ultimately a few thousand years of when layer were laid down in Earth's earlier history.

Being able to calculate the paleotemperatures and also the greenhouse gas, moisture, dust, and trace element content, is a researcher's dream. Unfortunately, there have been no cores taken from undisturbed ice-bearing localities for times earlier than 400,000 years, although an attempt is currently planned in Antarctica that may encounter ice that is a million years old. Let's hope it is undisturbed.

Without this record of the astonishingly rhythmic oscillations of the warm and cold periods every 100,000 years and the frequency and magnitude of the many smaller climate changes in between that correlate with Earth's tilt and wobble influences, the uproar over mankind causing all of the current "global warming" would be much louder. Since humans had no influence on these earlier cycles, such as contributing carbon dioxide or methane, it makes it more difficult to argue that human's input of CO_2 has become the primary driver of any climate change, a conclusion promoted or at least implied in most media stories today.

As mentioned earlier, the greatest shock that has come from detailed examination of these ice cores has been that CO_2 and methane changes followed or lagged the temperature changes. Another great shock that came from analyzing these cores was the fact that not only had the climate changed relatively often, but the climate had sometimes changed rapidly and the magnitude of the changes were frequently large. Some of the changes occurred in a few years to a few decades and changed mid-to-high latitude temperatures by many degrees Celsius. See Figure 32 for the record of temperature changes in Greenland over the last 120,000 years. A sudden change of this magnitude today could have a devastating effect on civilization as we know it. No one can possibly argue that these very large and rapid climate changes which dwarf any changes in the last 10,000 years were caused by humans.

Fig. 32

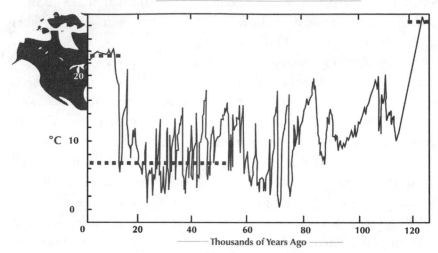

TEMPERATURE CHANGES IN GREENLAND

These temperature changes recorded in ice cores from central Greenland demonstrate the rapid climate shifts that have occurred during the last glacial stage and into the beginning of the current interglacial warm period (Paradise!) The average change (dotted lines)was about 16 degrees Celsius or 30 degrees Fahrenheit. While the change at lower latitudes would have been less, the colder changes and effects on growing seasons would have been disasterous. Note Earth's temperature at the end of the previous interglacial (upper right hand corner). Sea level was 19 feet higher than today.

Modified after L. Gerhard, 2001 AAPG Studies in Geology #47

Ocean Cores

Continual layers of information-bearing sediments can be found that go back to about 175 million years ago. Beyond that time, no sea floor record exists because all of the older sea floor has been subducted through the processes of plate tectonics, sea floor spreading, or continental drift (described in Chapter Three).

The information contained in these continuous layers includes fossils of plant and animal life that existed at that time, the amount and type of dust and the land source from which it came, inferred wind velocities, old glacial deposits that identify very cold periods in Earth's history, earlier ocean temperatures and chemistries, and sediment type and grain sizes that can infer whether the sediment was deposited by rivers near the coastlines or deposited in the ocean. The ocean cores are extremely important sources of an uninterrupted sequence of information on paleotemperatures, paleomarine environments, and paleoatmospheric content.

Lake & Bog Cores

The sediment layers found in old lakes and bogs contain plant and plant pollen remains, shells, animal remains, dust, and various sediment types. The analyses of these contents reveal information similar to that found in ocean cores except that the source of the sediment layers is more localized to the lake or bogs whereas some deep-sea sediment contents have been transported over long distances by submarine currents or drifting glaciers. The oldest cores from lakes or bogs normally date back only 25,000 years, although one bog deposit has yielded a core with sediments dating to 40,000 year ago.

Glacial Footprints

As continental glaciers began to melt at their lowest-latitude position, they deposited a lot of rock debris that had been ground up and incorporated into their base layers as they slowly moved down-slope. These southernmost positions (northern hemisphere), called terminal moraines, thus tell us of the extent of worldwide glacial coverage during that particular glacial period.

Where glaciers move to the coastline and "calve" or break off into the ocean, as they do whether the glaciers are either in a growing or melting phase, they dump a lot of debris where the then icebergs begin their melting in the ocean. When layers of such glacial debris, including boulder-sized pieces, are found on the sea floor at very low latitudes, a reasonable inference can be made that it was a very cold time in the ocean and atmosphere to have allowed so many of these icebergs to float to such low latitude without having already melted. During very cold periods, glaciers could have formed on mid-latitude landmasses themselves and those glaciers also leave footprints.

As the glaciers move slowly across the land surface, carrying all of this very abrasive rock debris, they leave recognizable scars or striations on the underlying solid rock formations. These footprints of glaciation are often the only visible record that remains of colder periods that occurred hundreds of millions of years ago, because, as mentioned earlier, the sea floor record goes back only 175 million years as a result of the subduction process, and much of the very old rock formations are buried under younger deposits. The oldest-known glacially created striations that have been found on the surface of rock formations are more than 2 billion years old.

Paleosols (Paleosoils)

Old soils are usually indicative of the climate in which they were formed. The soils contain the remains of plants, trees, insects, and minerals, and the general chemistry of the immediately underlying or adjacent rock or sediment. Simplistically put, soils created in cold climates contain remnants of cold plant types and dry climates contain assemblages of dry plant types, and so on.

Isotopic analyses of the content of these old soils often yield many clues needed for reliable paleoclimate reconstructions. An article in *Science* by Montanez, et al., 2007, contains detailed estimate of atmospheric CO_2, sea level, and also sea-surface temperature over a time from the late Carboniferous to the Early Permian. The more closely spaced data points reveal more detail than the smoothed curves presented in my generalized Grand Summary, which were based on 10-million-year sampling intervals, and provide some of the evidence that is missing in the generalized curves (see Figure 4, page 4 and Figure 33). As additional research is performed on the intermediate intervals, more details of the climate changes will be revealed.

Fig. 33

EARLY PERMIAN TEMPERATURE AND CO2 LEVELS

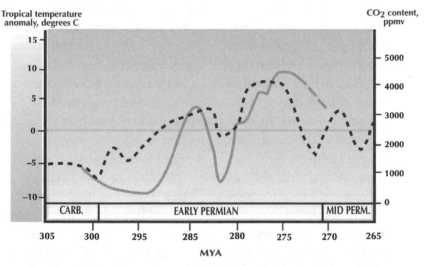

Inferred tropical temperatures (dashed line) and CO2 content (solid line) based on isotopic analysis of soil-formed minerals, fossil plant matter, and brachiopod shells. Only the midpoints of the ranges are shown.

After Montanez, et al., 2007

Fossil Assemblages

The observation that fossils of earlier life only persisted in a particular morphology or structure for finite periods of time was the first way rock assemblages were successfully correlated over significant distances of Earth's surface. The formations were judged to be time-equivalent if they contained identical fossils or fossil assemblages. Later, high-tech age-dating methods proved that this was indeed a proper application of the fossil information.

Since the animal and plant fossils changed over time in their morphology or structure, using fossils along with the superposition rule that younger sedimentary layers always overlie the older rocks beneath, allowed geologists to reconstruct the rock sequences. Unless the rock layers had been contorted by collisions between drifting continents or oceanic plates or other tectonic forces, these methods provided for the construction of reasonably accurate geologic maps.

The exceptions to the superposition rule were where younger volcanic, high-pressured, molten rock were injected into or between older rock formations or where thrust faults, that cause one side of a fault to move both horizontally and vertically, had pushed older rock along a fault plane and then up and on top of younger rock. This gave the early mappers a lot of trouble in understanding what they were seeing. Some very large layers of rocks have been folded completely over where the collision of continental or oceanic plates have compressed these layers into much shorter, contorted shapes.

As observations of fossils became more extensive and refined, it was found that many fossil types only occurred in a particular environment. For instance, planktonic foraminifera lived within the upper level of the oceans, and benthonic foraminifera lived at greater depths and at various levels near or in the bottom sediment of the oceans. Accordingly, if you found only planktonic foraminifera in the sediment layer, you could infer that the depositional environment was fairly shallow, since no deeper-dwelling benthonic forms were present. On land, certain beetles moved back and forth as the climate warmed and cooled, and some of the body parts of these temperature-sensitive insects were commonly preserved in the rocks, just waiting to tell a story about the old climate.

Rock Type

Almost all sedimentary rock gives an indication of the environment of deposition at the time they were deposited. While some river and stream deposits have similar depositional geometrics regardless of the climate in which they were deposited, other rock types, such as extensive reef formations, generally indicate deposition in shallow coastal waters or shallow inland seas that were located at mid-to-low latitude positions where the water temperatures were relatively warm. Evaporite formations containing salt, anhydrite, or gypsum deposits indicate an arid (hot and dry) environment of deposition.

Certain fossil assemblages that are present in the various rock formations give researchers additional hints as to what the climate may have been at the time the rocks were deposited. Also helpful are cores taken from wells drilled to find oil, gas, coal, or mineral deposits. Such cores can be observed and analyzed when brought to a laboratory and they can tell us the history of formations now buried thousands of feet below the surface of Earth.

Since, as you saw in Chapter Three, Earth's albedo or heat-absorbing potential has been identified as important in determining the climate, a sandstone found buried between some evaporate deposits conjures up images of a desert-like area of relatively high albedo that would have reflected a lot of sunlight, while a sandstone that contains tropical plant remains would be interpreted to be deposited from a river coursing through a low-latitude forest. This would be indicative of an area of very low albedo that would have absorbed most of the heat coming from the sun.

Remember that stable isotope studies of some of the minerals found within these sedimentary layers also provide information that tells us of the characteristics of earth's old atmosphere and climate.

Stalagmites and Stalactites

A somewhat unusual rock type is precipitated in caves when rain water or melted snow enters the ground and trickles down through cracks or permeable soils and drips from the ceiling of caves to form solid "rocks" called stalactites on the ceilings and stalagmites on the floors of the caves. The water trickling over the surface of the stalagmites and stalactites deposits some of the dissolved minerals the water has picked up on its journey to the cave in addition to the elements already in the raindrops. The isotropic ratios of oxygen and carbon provide a proxy record of the climate which the water or melted snow had been in when it entered the cave system. For instance, a high ratio of ^{18}O to ^{16}O would indicate a warm climate a low ratio of ^{18}O to ^{16}O a cold climate. The amount or rate of precipitation of the layers on the cave "rocks" and coloring indicative of organic content also would give clues of the wetness or aridity of the climate. These growing cave "rocks" can exist for long periods of time and can furnish valuable information from Earth's paleoclimate library.

Atmospheric Sampling

Today, we can measure the exact temperature and amount of the various greenhouse gases, oxygen, sulphur dioxide, other particulate matter, and moisture content of our atmosphere. While this is quite exciting, the time interval of the sampling has been so short that we haven't had time to observe any long-term trends to help us evaluate the causes and effects of precise changes in the atmosphere or their impact on the climate.

The very recent trends of rising temperature and CO_2 levels in the 1980s and 1990s do appear to track each other somewhat, but there were many times in the not-too-distant past when these trends did not tract each other (See Figure 19, page 32 and Figure 20, page 33). Keep in mind that many reputable scientists, including H. M. Fischer, et al., 1999, U. Siegenthaler, et al., 2005, and many others who have exhaustively studied most of the ice cores, **say Earth's temperature changed BEFORE the CO2 levels followed suit**. In fact, after temperature levels began to rise at the end of the Little Ice Age, around 1715, Earth has had some cool stretches that included the "Here comes another ice age" period of the 1940s to the 1970s when glaciers began to advance again in the northern hemisphere and the rising CO_2 levels did not track the temperature trend.

I am not advocating that scientists stop studying the present and past impact of CO_2 on climate change, just that they also study all other key factors that may be very important in climate change. Try to keep an open mind when someone says climate change is all driven by CO_2.

Always regard the source of the information. Give most credence to facts from researchers who publish in peer-reviewed journals such as *Science* and *Nature* and not those that shape their message in order to scare the money out of you. But even in these publications, you will find differing conclusions based on the same basic data. The science is not settled.

In the meantime, since we don't know all the answers, I recommend that we work toward lowering some of our greenhouse emissions. I just don't recommend a single focus on CO_2 to the detriment of studying other climate drivers or other more pressing problems that humanity now faces. More on this later.

Tree Rings, Pollen and Leaf Shapes

The variation in width of the growth rings of a tree can be used as a fairly good proxy for the general temperature and moisture record for a given year. Trees grow faster in warm, moist years and slower in cold, dry years or periods. Not surprisingly, tree rings also grow wider when CO_2 is elevated. The more trees you sample, the better the confidence in the outcome.

Unfortunately, trees don't live very long, a few thousand years at best. But if you can find some nearby buried trees that also live a long time, some of which overlapped the lifetime of your live tree, you can probe a little further into the past with some confidence.

For still older buried trees with no overlap, you can get an idea of their age back to about 50,000 years through carbon 14 age dating and tell by the configuration of the growth rings whether the climate was relatively stable and cold or relatively stable, warm, and moist or some fluctuations in between. Tree rings simply provide additional answer to another piece of the paleoclimate puzzle.

Fossil leaves or their imprints show up throughout the sedimentary record. The shape of leaf edges has been found to indicate whether the tree was growing in a warm or cold climate. Smooth margin leaves are found in the tropic, and jagged margin leaves are found in the colder climates. This correlation is quite strong and gives us still another clue to paleoclimates.

The hard case in which the fertilizing part of the pollen is contained often survives burial and becomes fossilized. These tiny cases can be identified and traced back to the types of plants from which they came. Since individual plant types usually grow within specific climate environments, the fossilized pollen are very useful in determining the general paleoclimate at the time the sediment containing the fossils was buried.

Corals

Like different plant species, some corals live only in very warm water, while other coral types can thrive in cooler waters. As scientists cross-correlate certain paleocoral types with other paleoclimate indicators, the confidence improves in using various corals as good indicators of regional paleoclimates or even warm or cold ocean current locations.

Organisms that build corals secrete calcareous skeletons or structures $(CaCo_3)$ so measurements can be made of the isotopic ratios of ^{18}O to ^{16}O, and ^{13}C to ^{12}C that can be used to estimate the climate at the time the coral was living. Thus, corals provide another indication of relative climate, and if enough samples can be obtained worldwide, researcher can make global paleoclimate estimates.

Most corals grow in a limited, shallow water depth range. This fact lets us use corals to establish old sea-level positions. Corals normally indicate a location along a shoreline or shallow shelf, although they also can be indicative of a location on an island or atoll. Very high average sea level positions indicate a time when a lot of water covered the interiors of the continents. These inland sea helped moderate the local climates and reduce the expanse of the deserts. Low average sea level positions often indicate a climate that should have been somewhat harsher and drier as you move inland from the coastlines.

Other factors, such as increased or decreased sea floor spreading rates or tectonic uplift or subsidence can also affect paleoshoreline positions independent of the climate at that time. As spreading rates increase and more hot, molten "rock" rises to or near the surface of the sea floor, the large and very long ocean ridges bulge upward and displace more seawater onto the continental margins. When spreading rates decline, the ridges cool, shrink, and subside, and sea level falls. Again, all pieces of the puzzle need to be examined before a reliable cause-and-effect analysis can be presented with some confidence.

Paleoshoreline Positions

As mentioned in the discussion on corals, paleoshoreline positions indicate the relative sea level at any given time, including the times when the shorelines moved off the continental shelve of the oceans and into the interiors of the continents. Knowing these shoreline positions, when combined with other indicator data, can help with a better interpretation of what was occurring at a particular time that may have had a directional influence on the climate. For instance, if sea levels were unusually high on a global scale but there was evidence of glaciations, the swelling of the mid-ocean ridges caused by accelerated sea floor spreading rates would be a likely culprit. This could suggest a time of more CO_2 being emitted into the atmosphere from the spreading centers or from the more active subduction zones and their related volcanic activity.

While a time of no glaciation also can cause high sea level stands, there is usually a limit on how high the melted ice volumes can cause the sea level to rise. Alternatively, very low sea level stands could indicate a time of very low mid-ocean spreading rates if little evidence of glaciation exists. Low spreading rates allow the magma extruded at the spreading center to cool and sink and not be replaced so quickly with a bulge of large volumes of more hot magma.

Thermometers, Written Records and Archeology

These indicators speak for themselves and are sufficiently covered in the Climate Indicators Summary in Figure 30, page 66.

Chapter Five

EXPANDING ON THE GEOLOGIC PERIODS

This chapter will offer some descriptions of the major geologic intervals that are shown on the Grand Summary Chart on page 4. As with the temperature, CO_2, and oxygen curves that are generalized because of the exceedingly long time between data-collection intervals, these observations are also generalized.

When more data becomes available, I am certain these curves and generalizations will be filled with the same constant climate variability that we now know has occurred in the more recent past and <u>is continuing</u> today. As you read these expanded summaries, you will find it helpful to frequently refer to the overall summary chart mentioned above.

PRE-CAMBRIAN EON

Archean Era, 4.6 billion to 2.5 billion years ago

The sun and Earth and other planets formed about 4.6 billion years ago. About 50 million years later, most astronomers believe a Mars-sized body collided with Earth, and the ensuing debris formed the moon. Earth's surface was initially molten, the sun's energy was 30 percent less than today, and the moon was much, much closer to Earth (less than 20,000 miles away) resulting in tremendous tidal effects on Earth's molten surface. As Earth gradually cooled, some solid rock began to form and clump into the earliest continents which floated on the molten surface. Space within the solar system was filled with debris, and these "rocks" of all sizes bombarded Earth very frequently which helped keep Earth's surface hot.

After about three-quarters of a billion years, the frequency of the extraterrestrial collisions dropped sharply, and Earth cooled at a more rapid pace. The early atmosphere, which contained no pure oxygen, was filled with carbon dioxide, nitrogen, and some water vapor and, before 3.8 billion years ago, probably saw water precipitation and the formation of early oceans. The tremendous tidal effects from the still-close moon, resulting in tides of several hundred feet, would have been unimaginable. Tides and waves would have been constantly eroding any early continents that had formed. Analyses of both oxygen and silicon isotopes have provided data that some researchers say indicate that ocean surface temperatures were quite high for much of the time between 3.5 and 2.5 billion years ago, but other researchers claim to have found evidence for glaciation (extreme ice cover) before the end of this era. The time intervals are so long that both may be correct.

A pivotal thing happened about 3.5 billion years ago with the occurrence of cynobacteria, the earliest form of life currently known from studying Earth's history. As this life flourished over the next one billion years, it used the sun's energy to sustain its own life and released oxygen into the atmosphere. That ultimately paved the way for the great oxidation event of 2.4 billion years ago. The very hot environment that gave rise to the cynobacteria still can be found in places on Earth's surface, such as in the thermal pools in Yellowstone Park, where cynobacteria still thrive in waters of 160 to 200 degrees Fahrenheit!

Best estimates, based on data from a few of the remaining ancient rock outcrops, suggest that several continents of modest size were present south of Earth's equator with three or four much smaller ones scattered elsewhere (see Figure 4, page 4).

Proterozoic Era, 2.5 billion to 540 million years ago

Many things of interest occurred during this next long era of Earth's history. The atmosphere finally contained significant oxygen, photosynthesis began, the first multi-celled algae, early aquatic plants, and some fungus appeared, and probably two periods of global glaciation occurred.

Best estimates of measurements of CO_2 and oxygen using stable isotope analyses indicate that CO_2 levels near the end of the era were almost twenty times current levels and oxygen content was slightly lower than today. Sea-surface temperatures, as indicated by stable isotope measurements, cooled to 20 to 30 degrees Celsius by 1.5 billion years ago. Current average near-surface global temperature is 15 degrees Celsius or 59 degrees Fahrenheit.

Cooling of Earth took a dramatic step around 850 million to 650 million years ago in that indications of significant glaciations, possibly three or more separate ones, show up in the geologic record. The length of the glaciations that left the characteristic scratches or striations on the rocks is not well known, however, and the CO_2 content and sea-surface temperature during or immediately prior to the glaciations is not well known, either. I assume lower sea-surface temperatures existed at the time of the glaciations!

Cooling of Earth also had helped create more solid landmass and a large continent had coalesced near the equator by around 2.0 billion to 1.5 billion years ago. By the time of the just-mentioned glaciations however, the large continent had broken into many smaller continents, all of which show some evidence of continental glaciation. Finding evidence of glaciation on all continents supports the idea that a large portion of the surface of Earth may have been encountering glaciation, which has given rise to what many refer to as the "Snowball Earth" period.

PALEOZOIC ERA

Cambrian Period, 540 million to 490 million years ago

Earth warmed again by the beginning of the Cambrian Period. The CO_2 levels were very high (18 times higher than today), and oxygen levels were also high, near to those oxygen values on Earth today (see Figure 4, page 4). Temperatures probably "stabilized" somewhat due, in part, to the greatly diminished frequency of extraterrestrial impacts.

The single most impressive event of all the eras lies in the explosion of life forms and the abundance of life that occurred during the

Cambrian. While some forms had begun to appear near the end of the Proterozoic Era such as worms, jellyfish, corals, and sponges, their abundance grew dramatically during the Cambrian, and new life forms appeared including clams, snails, squid, octopi, trilobites, insects, spiders, crabs, shrimp, and early fish. All of this happened in only a 50-million-year period whereas the earlier eras each covered a couple of billion years.

A very large continent, Gondwana, had formed in the Southern Hemisphere and was surrounded to the east, west, and north by much smaller continents.

Near the end of the Cambrian, oxygen and CO_2 levels dropped, and many of the new life forms in the sea became extinct, while others began to acquire a different morphology (internal or external body forms) to survive the atmospheric and upper-ocean chemistry change.

Ordovician Period, 490 million to 440 million years ago

The average temperature began quite warm, and CO_2 levels, while lower than in the Cambrian, were still high relative to succeeding periods. Oxygen levels dropped to very low values during the early and mid-Ordovician and then began a steady increase that would continue from more than 60 million years. The increase in oxygen possibly helped oxygenate the near surface oceans and stimulate a rebound of life that had an abundance of coral reefs, true fish, sponges, and other shelled organisms such as brachiopods and mollusks. Many other new species appeared.

The temperature declined until the Late Ordovician but still ended the period probably several degrees warmer than Earth's temperature today. Near the end of the period, the temperature began a protracted increase that would continue for more than 60 million years.

The Appalachian Mountains were being uplifted during the early and middle Ordovician, mountains that have since worn down dramatically from their earlier, near-Himalayan heights.

Gondwana began to drift toward and partially over the South Pole, with several smaller continents located in near-equatorial positions.

The Ordovician ended with a large extinction event, possibly from a major extraterrestrial impact.

Silurian Period, 440 million to 400 million years ago

While CO_2 levels dropped to the lowest levels up to that time, but still many times higher than today's levels, the oxygen content of the atmosphere rose to its then-highest level, which stable isotope calculations suggest were several percentage points higher than today's level. The moderately warm early Silurian temperature dropped for some relatively brief intervals into icehouse conditions around 440 to 430 million years ago, then warmed again. The very large continent, Gondwana, was still south of the equator and extended to the South Pole. Evidence of the glaciation exists in the rocks that were situated on Gondwana and some evidence indicates a glaciation extended to as low as 60 degrees south latitude, despite general CO_2 levels being up to ten times higher than today.

Life in the sea was very abundant and included huge eurypterids (scorpion like creatures ten feet long that ruled the seas).

The major life-changing events of the Silurian, however, took place not in the sea but on land. Life moved from the sea to the land and this included the first land plants. The surface of Earth began to look green for the first time! Just imagine, for more than four billion years Earth's land surface had no plants or animals! The first record of animal life on land was that of scorpion, insect, and spider fossils.

The very high levels of oxygen must have aided the growth of much of the animal life because many of the scorpions and other insects were huge, much larger than when they lived in times of lower oxygen.

High concentrations of oxygen usually mean a high incidence of fire. The earliest land plants probably paid a price for finally making landfall during this time of elevated oxygen. Evidence of fire is indicated by the presence of large amounts of charcoal in the rocks of this time.

Devonian Period, 400 million to 360 million years ago

The Devonian began with moderate and rising temperature and CO_2 levels, which, by the Middle Devonian, began a long, significant decline that continued into the next geologic period. Oxygen levels, which were very high, declined dramatically during the Early Devonian, bottoming out in the Mid-Devonian at the lowest level since the Pre-Cambrian Eon. This dramatic decline, although rising somewhat in the Late Devonian, may have combined with extraterrestrial impact activity at the end of the period to cause the large extinction event that occurred at that time (See Figure 4, page 4).

The first amphibians appeared during the Devonian which meant that, for the first time, non-insect animals that had evolved from the fish had finally made landfall. Also, some flying insects made their appearance. This transition is recorded in the fossil record of that time. The first true sharks and rays had appeared in the seas and they survived this large extinction event and all that have occurred since. As far as I know, no one knows for sure why sharks were more resilient than most of the other life forms present at that and subsequent times.

Gondwana still occupied the Southern Hemisphere and several of the smaller continental landmasses had begun to clump together slowly near the equator.

Land plants were continuing to expand, and many mosses, rushes, and ferns appeared.

Carboniferous Period, 360 million to 300 million years ago

The Carboniferous began warm and wet, but the temperature decline set in motion in the Devonian Period continued and helped change the fauna and flora before the end of the Carboniferous. The first trees appeared, and the vegetation was lush in the widespread swamps around the world, probably helping to cause the plunge in CO_2. All of this vegetation ultimately would be buried in these vast swamps and form 90 percent of the coal deposits present in the world today. Despite the extensive swamps that had occupied large portions of the now present two huge continents, Gondwana and Laurasia, they both also contained very large deserts in their interiors which is a common feature where land is located so far from a coastal area.

Toward the end of the Early Carboniferous, oxygen levels began a dramatic rise. Flying insects appeared that had wing spans of up to 3 feet for the paleo-dragonflies. Then came the first reptiles. The fossil record indicates that huge forest fires occurred at this time, probably triggered by the high and rising oxygen levels.

The two huge continents, Gondwana and Laurasia, portions of which were located over the equator, probably helped divert equatorial currents to higher latitudes and caused or aided the onset of glaciation by providing more moisture to the coldest areas of southernmost Gondwana.

Global temperature continued to drop throughout the Carboniferous, and by the Middle Carboniferous, about 325 million years ago, the CO_2 content of the atmosphere plummeted below any currently identified earlier levels and almost as low as today's level. This icehouse cycle, with continuing glaciations, persisted for the remainder of the period and throughout most of the next period, the Permian.

Permian Period, 300 million to 250 million years ago

The very low CO_2 levels of the Late Carboniferous persisted into the Permian. The icehouse (glacial) conditions also continued throughout most of the Permian. Oxygen levels soared to approximately 30 percent, which is the highest level measured in the history of Earth. That level would be about 50 percent higher than today's level. Even you and I might have been able to finish a 26-mile marathon. Near the end of the Permian, oxygen levels plummeted. Temperatures recovered to modestly high levels, much warmer than today, and the icehouse period ended.

The huge continents present during the previous period had collided and formed one supercontinent, called Pangaea. For the first time, all life forms that had evolved on land could "communicate" with one another, absent an unclimbable mountain range or impenetrable desert barrier. The deserts were very vast and that evidence shows up in the geologic record in the form of thick evaporite deposits such as salt, gypsum, and anhydrite.

Life on Earth diversified further with lizards, snakes, and some animal-like creatures, the synapsida, appearing. The best known of the synapsida were the Dimetrodon and its other four-legged, air breathing cousins. Forerunners of the dinosaurs appeared near the end of the period. All in all, this period of very high oxygen levels did not see as much diversification as other periods that were more oxygen-starved. This might suggest that times of oxygen starvation created pressure for the animal kingdom to diversify in order to survive (Ward, P. D., 2006). Meanwhile, the long-term, low-levels of CO_2 were not conducive to robust plant life on this supercontinent.

At the end of the Permian, all hell broke loose. Catastrophe, or whatever superlative you want to call it, caused approximately 90 percent of all life on land and in the sea to die in a geologic instant of time. Oxygen levels had crashed and CO_2 levels had risen. Other things happened, causative things responsible for this greatest of all extinctions. Most scientists believe a giant asteroid struck Earth in the vicinity of present-day Antarctica, which was then the continent of Australia that was part of the supercontinent called Pangaea. As mentioned earlier, researchers from Ohio State University have recently located a gravity anomaly that has the earmarks of a supergiant impact crater about 300 miles wide beneath the eastern ice shelf of Antarctica. The crater that wiped out the dinosaurs 65 million years ago was "only" 110 miles wide. If the crater is from a giant impact of about 250 million years ago, as is claimed in the initial estimates, the researchers have found the cause of the greatest loss of life in Earth's history.

Concurrently, a vast volcanic field erupted in Siberia that ultimately covered an area the size of Alaska. Either of the events would have killed off a huge amount of life, but having both happen simultaneously would have just made things worse.

Some scientists say there was no extraterrestrial impact, that the volcanic eruptions did it all. I don't disagree that super-volcanic eruptions over a long period of time could have been the killers by causing acid rain, but why the coincidence of other mega volcanic eruptions with these giant impacts? The timing of the arrivals of large asteroids is certainly not caused by volcanic eruptions!

My guess is that both things happened. The shock waves from the impact by an asteroid estimated to have been 30 miles wide, which was many multiples larger in total mass than the one that supposedly killed off the dinosaurs, rolled around the surface of Earth and/or through Earth's interior and caused such violent earthquake and fissure in the outer crust that the Siberian Traps (volcanic field) on the opposite side of Earth sprang to life or, if already present in a smaller area, probably expanded immediately with the newly formed lava conduits. Any faults that had pent-up tensions would have had an opportunity to release those tensions as a result of this impact-caused ground roll.

The atmosphere would have been full of toxic acid rain and toxic to life. The oceans would have been poisoned except for the deeper environs. How anything on the surface survived is a mystery in itself. But some did survive! Let's go to the next period and find out not only what survived but also what ultimately flourished.

Triassic Period, 250 to 200 million years ago

Oxygen levels were relatively low through out the Triassic. CO_2 levels began to fall again following the extinction event but apparently "stabilized" at a modestly low level. Temperatures were high early but dropped significantly near the end of the Triassic. Temperatures were, however, still warmer than those today.

The near-lifeless Earth persisted for a few million years, and then, around 240 million years ago, life began to recover and transform into species that could breathe adequately in the low-oxygen environment. The new life forms flourished, helped by the fact that little competition existed.

In the seas, many new species developed, while many other were modified. New types of coral, mollusks, ammonites, and sea creatures (ichthyosaurs) appeared in the recovering seas. To see what some of these early creatures probably looked like watch some of the programs that cover these ancient periods on the *Science*, *History*, or *Discovery* channels.

On land, the few forerunners of the dinosaurs developed into multiple combinations of body sizes and shapes. Some of these remnants of the Permian extinction took to the air and became the first flying reptiles (pterosaurs). Many large mammal-like creatures existed. But none was a match for the two-legged dinosaurs that had developed a breathing system that allowed them to utilize the small amount of oxygen much more efficiently (Ward, 2006) and the 185-million-year reign of the dinosaur began. Meanwhile, the earliest true mammals evolved and, because of their diminutive size, escaped extinction at the hands of the carnivorous dinosaurs.

Pangaea, the supercontinent, was initially intact but was beginning to show signs of breaking up as the period ended. A few asteroids of some size struck Earth in the Late Triassic but no large climate change has yet been identified with these asteroids impacts.

Another mass extinction event occurred at the end of the Triassic, possibly caused by a further reduction in the oxygen content of the atmosphere.

Jurassic Period, 200 million to 145 million years ago

In the Jurassic, oxygen levels were at the lowest levels recorded in at least the last 500 million years. The early low temperature increased slightly by the latter half of the Jurassic, and so had the CO_2 level.

For the larger animals present, the low-oxygen environment continued to favor the dinosaurs. While mammal species expanded in numbers, they remained quite small in size. Some of the dinosaurs that had developed feathers in the Triassic continued evolving into birds. Other dinosaurs became true giants by the Late Jurassic.

Life in the sea, which had been affected by the Triassic extinction event, recovered well and appeared to have less trouble living in a lower-oxygen world than did the larger, non-dinosaur creatures on land. All of the sea life with calcareous (limestone-like) shells flourished in a higher CO_2 environment because it was partly from the CO_2 that the creatures derived their carbonate shells. The CO_2 level, many times higher than today's, apparently did not "acidify" the oceans enough to prevent the formation of abundant shells for life in the sea, as some scientists fear may happen with a modest increase in CO_2 levels in today's seas. However, it is possible that more detailed sampling of the Jurassic sediment may find that shell abundance and CO_2 levels did generally vary together.

Pangaea began to break up during the Jurassic, and this included the early drifting apart of South America from Africa and North America from Europe. The major environmental impact from this break-up would have been the reduction in size of the great deserts, since the newly formed waterways brought moisture to the interior of the fracturing Pangaea.

Cretaceous Period, 145 million to 65 million years ago

Moving nearer to the present, much more data is available. Subduction during continental drift has not erased all of the continuous record of the sea floor sediments as it has for sediments that were older than 175 millions years of age.

The Cretaceous atmosphere began with a high CO_2 level, moderate temperatures and a humid climate. While the CO_2 level declined later in the period, it was about 10 times higher in the Early Cretaceous than today's interglacial average. Several estimates place the peak global temperature at about 20 degrees Fahrenheit higher than at present! The oxygen level started out very low but rose throughout the period, concurrent with the reduction in CO_2 levels.

Sea floor spreading rates were much higher than at present, and sea level was at least 600 or 700 feet higher than today. Shallow seas covered the interior of many continents, including the central interior of North America. Marine life flourished and corals and rudistids grew in abundance. The rudistids, mollusk (clam-like) animals that attached themselves together to form large reef-like structures, dominated the shallow seaways. Outcrops of Cretaceous rocks are full of either macrofossils that you can easily see such as corals and clams, or microfossils that are present in lime mudstones and chalks such as Cocoliths. The famous White Cliffs of Dover are composed of such chalk.

Although the dinosaurs didn't really need the rising oxygen level, they adapted well to it and continued to kick butt in the chow line.

The continents were well scattered and not too dissimilar from today's distribution except that North and South America were much closer to Europe and Africa. No major continental collisions were occurring, no extensive glaciation existed, and, except for the rising oxygen level, the climate and atmosphere were probably about as stable as it gets.

Then nearly all hell broke loose again. An asteroid about six miles across and traveling at 40,000 to 60,000 miles per hour traversed Earth's atmosphere in about 3 or 4 seconds and struck shallow water or land at what is now the Yucatan Peninsula in Mexico. No doubt about this one; it left a crater that measures 110 miles wide! It is estimated to have created an initial wave 3,000 to 5,000 feet high which was still huge as it washed over a slightly submerged Florida, carrying boulders from the impact site out into the western Atlantic. Some tsunami!

Molten debris was lofted into space and fell all around Earth, setting fires worldwide and causing acid rain. Wind with velocities estimated at 600 miles per hour probably blew away huge dinosaurs unlucky enough to be within a hundred miles of ground zero. Just think what havoc the tsunami that swept up the interior of North America caused. The dust, which stayed in the atmosphere for years, darkened the sky and reflected most of the heat coming from the sun above. This turned the previously warm climate into an instant, albeit brief, icehouse.

About this time, the Deccan Traps volcanic field in India became active and expelled great volumes of magma and related gases. This occurrence exacerbated the toxicity of Earth's atmosphere and likely caused many more extinctions.

The dinosaurs and more than 60 percent of all species on Earth did not survive. Bad for them, lucky for us, because, as you will see in the next period, it gave rise to many more mammals, and eventually our ancestors showed up.

Tertiary Period, 65 million to 2.6 million years ago

The carbonate, limestone-rich, coral-rich, and dinosaur-ruled world of the Cretaceous changed in that fiery instant to a world of mostly clastic (sand) type sediments that were eroding from the suddenly more barren land surface. Many of the plants kept a toehold, however, and soon the world turned green again. The giant sequoias survived since they were growing in the shadow and protection of the mountains that exited in the west. New pine trees proliferated, as did cactus-type plants.

Most of the fish and many of the shellfish made it through the catastrophe with only a temporary reduction in numbers. The alligators and turtles, close relatives of the dinosaurs, survived essentially intact because both were able to go underground and hibernate for an extended time and sleep through the brief icehouse. When an alligator goes into hibernation, its heart rate drops from thirty beats per minute to two beats per minute and little oxygen is required.

After a few million years, two groups not only survived but did so in a big way. Birds evolved into many different types, and the mammals, heretofore small and rather unimpressive, diversified into many new species and became the dominant force on land. Many animals became huge and evolved into today's rhinos, elephants, hippos, and whales. Others species grew larger, more efficient brains, such as the monkeys, apes, and dolphins. Man's best friend's, the dogs and also the cats (some very large), proliferated, although with characteristics probably much wilder and more aggressive than today's household varieties.

The consensus of anthropologists is that our ancestors, the *Homo sapiens*, slowly evolved from the Tertiary Period apes over a period of several million years. By God, intelligent design, just plain evolution or some combination, our brain sizes increased and developed the ability to reason and plan and be aware of self and others and of how the actions we take today will likely affect tomorrow. We have no peers and are clearly the most dominant force Earth has ever known, even though we lack the claws or teeth of the earlier rulers of their time. The long-in-coming paradise that we live in today is only spoiled by our inability to get along with each other. So we have invented things much more formidable than large claws and teeth. What a pity.

The climate changed a lot during the Tertiary period which ended 2.6 million years ago. Earth's temperature continued a long decline accompanied by lower and lower CO_2 values (see Figure 24, page 48). A couple of the brief exceptions were the CO_2 spikes following the dinosaur-killing asteroid impact of 65 million years ago and the significant temperature rise 55 million years ago which is, interestingly, near the beginning of the collision of the Indian plate with the Asian plate. The long-term drop in atmospheric CO_2 may have been caused or at least aided by the sudden increase in chemical weathering brought about by this collision which caused a great increase in elevation of the India-Asian contact area.

The sudden warm spike of 55 million years ago is thought by many to have been caused by a large methane (clathrate) release that caused sudden greenhouse warming. But methane is believed to only remain in the atmosphere about 10 years. Maybe the sudden warming started a chain reaction of positive feedback for warming. Maybe the solar-orbital factors were poised for a shift to a warmer period on Earth, and the methane burst just pulled the trigger. Maybe the sudden and rapid movement of India from near the South Pole to near its location today, caused great rates of subduction and disturbed large deposits of methane frozen in the sea floor. I'm sure the answer is to be found somewhere in Earth's library.

Then, about 35 million year ago, Earth began to cool more rapidly and the South Pole that was located atop the continent of Antarctica developed permanent glacial ice. By seven million years ago, Greenland had developed huge volumes of ice that persisted through the end of the Tertiary and remain there today.

With the exception of the ongoing collision of India with Asia, the rest of the continents were drifting rather freely on Earth's molten subsurface. How important was this one collision from a climate standpoint? High, fragile, rocky surfaces like those in the Himalayas and the Tibetan plateau were providing a constant new source of silicate rock fragments and surfaces that could undergo chemical weathering and lower, or continue to lower, the long-term CO_2 content of the atmosphere. As the CO_2 levels dropped, so did the temperatures . . . or was it vice-versa (see Figure 24, page 48)?

Quaternary Period, 2.6 million years ago to present

By the beginning of the Quaternary Period, 2.6 million years ago, very large ice sheets were present in the higher latitude of the Northern Hemisphere. The advance and retreat of these ice sheets have culminated in our current state of approximately 90,000 years of cold, windy, and dry glacial periods followed by rather brief 10,000-year interglacial, warmer, and wetter periods. These warm interludes are then followed by another 90,000 years of cold, windy, and dry glacial conditions.

Three or four million years ago, the open passageway for equatorial ocean currents between the Atlantic and Pacific Oceans was closed off by tectonic uplift and volcanic outpourings. This newly formed land is now the Isthmus of Panama. With the free movement of warm, low-latitude Atlantic waters "suddenly" blocked, the climate patterns began to change and for the first time in hundreds of millions of years, very large continental ice sheets began to form in the Northern Hemisphere. While Earth had experienced glacial conditions in the Southern Hemisphere over the previous 30 plus million years, the glacial cycle became more frequent and rhythmical in the Northern Hemisphere after the Panama closing.

The timing of the advances and retreats changed during the Quaternary Period. The earlier cycles of glacial to brief interglacial climates were around 41,000 years, and then, about 900,000 years ago, the cycles became about 100,000 years in length, and the ice sheet became even larger. The most recent glacial advance, which began to retreat only around 15,000 years ago, caused ice (from snow) to build up to nearly two miles thick just north of what is now New York (see Figure 34). These facts are real and not being debated. They are, of course, fodder for catastrophe-oriented movies and TV programs. While these ice buildups and relatively rapid switches to melting phases did occur, the melting and rising sea levels took thousand of years, not the instant change depicted in the movies or some television "documentaries".

What did occur very quickly, in a decade or less, was a flip of the climate from very cold to suddenly warm or to a somewhat lesser extent, vice-versa. Fortunately for our ancestors and relative to our future, the flip to the initial cold periods did not immediately take the climate to the depths of the glacial cycle temperatures. While that is true, as verified by careful studies of ice cores taken in Greenland, Antarctica, and from mountain glaciers, the temperature did "immediately" drop enough to cause instant havoc to the growing cycles in what are now the mid-latitude bread baskets of the world. From feast to worldwide famine in a few short years is a truly horrible scenario.

Let me cite some real examples of how much even small changes in global temperatures can impact Earth's civilizations. Around 1,350 A.D., the warmth of the Medieval Warm Period began to descend into a cooler and drier cycle known as the Little Ice Age. This cooling had an effect on global average temperature of only two to four degrees Fahrenheit. This change would have been even less near the equator but would have been still greater in the high latitudes. The Vikings had made it to North America and established villages surrounded by some agriculture and cattle ranching. But when this rather modest global climate cooling occurred, even this hearty group of explorers froze, starved to death, or blended in with the natives.

The Anazazi Indians, with their fabulous cliff dwellings in northern New Mexico, also disappeared "overnight" during this Little Ice Age and the attendant drought that affected their region. They may have simply moved south, but either the drying and dying of their croplands or the harsher weather was probably what caused them to abandon such fine dwellings.

In Central America, a strong case can be made that the very advanced and unusually long-lived civilization of the Mayans fell on hard times and disintegrated when a climate-related, prolonged drought changed their environment around 800 A.D. This probably forced them to abandon their marvelous cities during the next century.

Fig. 34

GLACIAL ADVANCES AND RETREATS

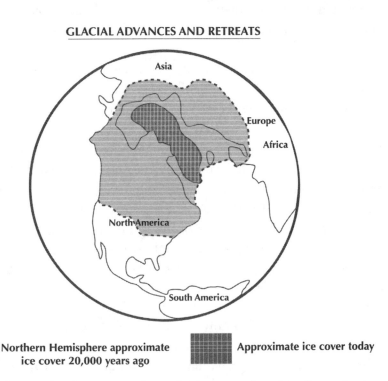

Northern Hemisphere approximate ice cover 20,000 years ago

Approximate ice cover today

As the archeologist, anthropologists, and paleoclimatologists continue to get together to review earlier civilizations, they are finding many more worldwide examples of the negative or positive effects of climate change on earlier civilizations. In nearly all cases, the dislocations were caused by colder and/or drier climates, not warmer and wetter climates like the one we are living in today.

With the exception of the brief and modest spikes in CO_2 during the interglacial cycles, worldwide CO_2 levels are as low as they have ever been measured over the span of Earth's history. So what? I know I'm really enjoying this brief paradise of a warm trend in which we are currently living and I certainly do not want the CO_2 levels to drop again below their pre-industrial levels of 280 *ppm*, especially if they are following the temperature down!

So here we are, basking in one of the very few instances of earthly paradise and environment for mankind. Walk outside and rejoice in its hospitality, whether it is zero or 100 degrees. Over the year, it averages about 59 degrees Fahrenheit worldwide and we can live very comfortably in the temperatures and growing seasons that 59 degrees Fahrenheit provide.

Chapter Six

MORE ON THE GREENHOUSE

Greenhouse gases are good!

Without greenhouse gases Earth would be frozen solid, its global average temperature approximately minus 18 degrees Celsius or zero degrees Fahrenheit. We are currently in a paradise of a greenhouse mixture and orbital orientation for humanity with this average global temperature of 59 degrees Fahrenheit.

How much variation in greenhouse gases can we tolerate and still be considered to be in a paradise mode? We don't know for sure. We do know that during the most recent glacial extremes, with CO_2 levels of only 180 *ppm* (currently 380 *ppm*) and methane volumes nearly half what they are today, that Earth was relatively uninhabitable north of about 40 degrees latitude in North America and Europe because of pervasive ice cover. A large number of Earth's people live near or above that latitude today; this includes major cities such as New York, Chicago, Seattle, Paris, London and Moscow. But since we now know that CO_2 levels follow, not cause, the temperature changes and that CO_2's ability to trap more heat declines for each additional molecule of CO_2 added to the atmosphere, the effect of rising CO_2 certainly appears not to be as significant as once thought by nearly all of us and still thought by most of the general population.

If Earth re-enters a cooling phase and the cooler oceans quickly absorb a lot of atmospheric CO_2, the greenhouse effect would decrease at a greater rate with each molecule of CO_2 decline. To say it another way, the ability of the next declining unit of CO_2 to cause more global greenhouse cooling is greater than the next increasing unit of CO_2's ability to cause more global greenhouse warming. This is because of the logarithmic decline properties of atmosphere CO_2.

Another detrimental effect of returning to the very low CO_2 levels of the glacial cycles would be the drop in global plant growth, not just from the cooler climate and attendant shrinkage of the agricultural growth belts, but in starving the flora of its life-giving CO_2. Hence, the additional potential benefit of sequestering CO_2 today in case it is needed to try and counter the many effects of global cooling as CO_2 levels become very low again.

During the Cretaceous Period, when CO_2 saturations were 4 to 10 times higher than today's levels, humans were not around to have their survival skills tested. These much warmer climates did allow certain species of land and sea life to flourish however. It must have been particularly favorable for the growth of creatures with big teeth and claws! This super-hothouse period occurred 90 to 130 million years ago, when dinosaurs still roamed earth. Measurements of stable isotopes from that period indicate a probable variation from today's worldwide average annual temperature of 59 degrees Fahrenheit, of as much as 20 degrees Fahrenheit or a world wide average day and nighttime temperature of nearly 80 degrees Fahrenheit. Whew! It is little wonder that we find dinosaur fossils in rocks of this age far north of the Arctic Circle (66.7 degrees north latitude).

If we had been present during the Cretaceous and had not been eaten by the big lizards or sea monsters, we might have been able to survive but, except for the ice-free poles, would probably have always been hotter than blue blazes and feeling like we were breathing at 15,000 feet because of the lower oxygen content of the atmosphere (see Figure 4, page 4).

Was it just the high CO_2 level that created this hothouse period? It could have been, but additional research may find that this part of the Cretaceous was a time of very high solar activity that increased the sun's radiance somewhat or that the sun's magnetic activity spiked and created much more solar wind that shielded cosmic rays from Earth which possibly resulted in much less low cloud cover. Scientists do know the sea floor spreading rates were much higher during the

Cretaceous, so more heat from within Earth would have been delivered to the oceans. Or was it some other factor we have not yet recognized? Our descendents will find the answers in Earth's library.

If our climate were to move toward the warmer and wetter than today mode, conditions might not be too difficult to tolerate, since, generally, "warmer and wetter, most life lives better". A move in the opposite direction; "cold and dry, many life forms die".

Would diseases in a much warmer, wetter climate also have flourished? Possibly, unless the greater abundance of plants for food caused by the CO_2 enrichment, provided better nutrition and therefore resistance to diseases of the day. Disease did not appear to end the rule of the dinosaurs. Currently, the most scientifically accepted hypothesis is that the dino killers were a giant meteor impact and the accompanying large volcanic eruptions.

In summary, a global climate moderately warmer and wetter is certainly easier to tolerate and survive in than a moderately colder, drier, and windier climate. The world's agricultural output certainly would expand, both in more viable agricultural land at higher latitudes and growth enhancement because of the presence of more CO_2 (Figure 18, page 31). The historical distribution and abundance of plant and terrestrial animal life appear to support the fact that most life on Earth thrived during warmer, rather can cooler, climates. We have seen that mankind fared much better in the warmer times of the last 2,000 years. Having climate remain right where it is would be mighty fine, but based on history, an absolutely stable climate would be the most unusual climate event that has ever been documented.

While keeping Earth's temperature near its present value might seem ideal, we do not have the technology, in this first decade of the twenty-first century, to be able to "lock in" our current climate. It will change even if we do not tamper with it or if we should get CO_2 back down to the pre-industrial level of 280 *ppm*. We have no riskless incentive to push it artificially toward a warmer state or colder state. We need to study the information in Earth's library to see if it will help us get prepared to deal with whatever climate changes may occur, one way or the other, and eventually be able to strongly influence some of the climate drivers. Same answer, same recommendation; more research.

Chapter Seven

GENERALIZATIONS AND SCENARIO DISCUSSIONS

As the climate cools, global winds get stronger (and drier).
As the climate warms, global winds get less strong (and wetter).

Ice cores bear out these generalizations, but anomalies in the above are commonly thought to be hurricanes and cyclones that derive their energy from the warm surface waters of the oceans. In a general global warming, the average, lighter surface winds could get interrupted by occasional severe storms fueled by these warm surface waters. Recently, however, it has been found that high-velocity, high-altitude winds, during periods of warmer or cooler temperatures, can shear off the tops or tilt over the rotating storms and decrease their intensity. A study by members of Woods Hole Oceanographic Institute that analyzed 5,000 years of sediments in a normally isolated Caribbean lagoon (except during storms surges) has found that the hurricane frequency and intensity were just as high during periods when the temperatures were cooler. In fact, they found that intense hurricanes even made landfall during the latter half of the Little Ice Age. They further found that the times of more and less hurricane activity seemed to run in cycles that were not connected to temperatures in the Atlantic region. So, we must be careful about jumping to conclusions without examining all available data. Results of this study that were reported in Space and Science, 2007, point once again to the complexity of the climate system.

As a couple of scenarios are played out below, keep in mind that they are just that - scenarios. Although scenarios are simply possible outcomes under varying conditions, they help us do broader analyses of what the future might hold, as well as help us understand what might have happened in the past.

Scenario one; the current warming being experienced on Earth will not necessarily lead to a long-term, severe hothouse, but depending on the interaction of the many drivers or dominance by a key one, it could lead to such an outcome. For instance, as Earth begins to heat, the average wind velocity should drop, which would mean fewer nutrients would be blown from land to the open oceans. Diminishing the food available to life in the oceans that takes CO_2 from the atmosphere to make shells or skeletons would mean that less CO_2 is taken from the atmosphere and sequestered at the bottom of the ocean thus allowing a build-up of CO_2 in the ocean. Also, lighter winds would cause less mechanical upwelling of nutrient-rich waters that normally stimulate coastal organic blooms that result in removal of CO_2 from the atmosphere. Note here that I am saying "could" a lot and not "will" because there are other, more powerful drivers constantly at work. As the surface of the oceans warms, more CO_2 would be released to the atmosphere, increasing the greenhouse effect to some degree.

As CO_2 increased and the vegetation responded with accelerated growth, more methane would be created from the rotting plants, causing the addition of this other greenhouse gas. Also, possibly of even more importance, adding vegetation to cover former highly-reflective, barren land would lower Earth's albedo and result in more solar heat being absorbed at Earth's surface. With continued warming and rising sea levels caused by thermal expansion of the oceans and melting glaciers, sandy coastal plains would get covered with salt water of lower albedo, and covering former permafrost coasts with water also would mean lower albedo. Melting of some of the permafrost would release more methane and could result in more positive feedback for warming.

Polar and mountain glacial melting would mean that more ice-free low-albedo land would be exposed, which could mean still more positive feedback for warming. However, exposing previously buried rocks and soil would cause more chemical weathering and long-term removal of CO_2. But as the higher CO_2 levels that had been created in the short term stimulated robust plant growth, the lower albedo caused by the greening of land should overpower the much longer-term effect of chemically removing CO_2 from the atmosphere.

In the scenario above, with so many feedbacks apparently lined up to favor still more warming, what could possibly happen to save us from such a hellish future? Let's grab some facts, add some speculation, and play out an opposite scenario.

Scenario two; that the orbital alignment is gradually moving toward a cooling effect has not been disputed as far as I know. Some say the man-made CO_2 increase may overpower the orbital factor, but I personally doubt it, since CO_2's ability to trap more heat is probably near its effective maximum regarding temperature levels. So, let's continue. The current warming should provide an atmosphere that can, per the laws of physics, hold more water vapor. A higher temperature should cause more evaporation, which can feed the warmer atmosphere with more water vapor. A high level of water vapor (or moisture) in the air generally aids the development of more low-level clouds which block (reflect) some of the sun's rays and also lead to more precipitation, both of which promote cooling. However, since water vapor is by far the most dominant greenhouse gas, the argument can be made that the addition of more water vapor will overpower the cooling effects and lead to more warming. The answer to this question is not yet resolved. But to continue our scenario discussion, let's assume, as many scientists believe, that the cooling influence will win.

As Earth begins to cool, many of the drivers begin to add positive feedback for more cooling such as ice and snow and lower sea levels for increased albedo. Less vegetation in a cooler climate would mean less methane release if the shocking recent study is confirmed that LIVE plants may be releasing 10 percent to 40 percent of the greenhouse gas, methane, into the atmosphere. Even if there is simply less vegetation, there would be less to rot in the water where the methane is routinely formed and more precipitation generally means more chemical weathering, and that would remove CO_2 from the atmosphere at a higher rate. Less atmospheric CO_2 would mean less CO_2 to be utilized by the plants and thus fewer plants covering Earth's surface which would result in a higher albedo.

The suddenly cooler oceans would become a "sink" for removing large quantities of CO_2 from the atmosphere, which should cause less greenhouse effect and cause further cooling. The volumetric shrinking through simply cooling the oceans would help lower sea level by a small amount which would increase Earth's albedo somewhat by exposing more high-albedo shorelines.

On top of all of the above scenarios lies a factor that cannot be reliably predicted for the long term. That factor is the total amount of heat and magnetically generated activity and effects coming from the sun. The sun is Earth's ultimate heat engine. The sun's intensity does vary. Proof of that lies in the records of several hundred years of very lower sunspot and magnetic activity, called the Maunder Minimum, which, as you will remember, was coincident with the Little Ice Age. Having previously reviewed both real and potential solar effects, you now can grab your pencil and play out several additional warmer-climate or colder-climate scenarios.

It fascinates me how much time and fanfare we give to the greenhouse gas CO_2 and how little recognition is given to variations within the sun or with ocean currents. In fact, most of the global-warming modelers do not even include solar variations or include things such as the cloud-generating potential of cosmic rays in their models. While we cannot do anything about solar variations, we at least can acknowledge the possible contributions of the sun's effects when trying to explain historical global climate changes. We also should pay more attention to the forecasts of future solar cycles so that we might better prepare for solar-generated or solar-enhancing effects on the climate.

You now know the key factors that have been recognized to drive or affect the climate. Have some fun, and make your own predictions. But be careful not to make your prediction by considering the effect of changing only one of the climate drivers and thinking all of the other drivers will temporarily cease to exert their influences. They all keep moving and at different rates!

Chapter Eight

THE FUTURE

As you build your own vision of the future or simply follow the researchers' predictions, there will be flaws, certainly in detail. But as research continues and new data flood in to fill the gaps in the summaries gathered here, better-thought-out predictions should emerge that are closer to reality. Regard the source of the new data. Beware of summaries by the media that are designed to sell fear because the media knows alarming news sells better, even in a climate of paradise (shame on them!). Be aware of people who must squeak to get "grease" for their cause. It is human nature to do so. That is an area where the peer review boards should step up and identify some of the purposefully biased reports that appear in technical journals written by people on either extreme of the global-warming issue.

In making your prediction of the future, consider that sea level was nearly 20 feet higher at the end of the last interglacial cycle about 100,000 years ago. It got there without human's input of greenhouse gases. The temperature was about five degrees Celsius warmer than our current temperature. Our sea level could someday rise that much, even more. But it will not rise that much in just a century if it approaches that height at all.

Our current interval of general interglacial warmth may possibly last a few thousand more years because of natural, well-established forces. The orbital alignments that brought us out of the last ice age change very slowly, and this current period of warmth may have some residual orbital influences regarding continued warming. Humanity is changing the atmospheric mix of greenhouse gases that affect climate. Although of a debatable magnitude, most scientists originally believed the influence has been to enhance overall warming. Many now believe the human-induced increase in greenhouse gas to cause warming is negligible. Others reject that thought completely.

Regarding future cooling, we do not know at what point the slowly changing orbital influences will take over and start moving Earth in a cooling direction. Recent temperature measurements by satellites and the Hadley Climate Forecasting Center in the United Kingdom indicate that since year-end 2001, Earth's temperature has begun to trend downward from its rather constant recent ascent (Figure 20, page 33). Various verbal and written reports from Argentina, Australia, and China have described cooling trends for the last four to eight years. Does this indicate a new trend toward temporary, relative stability or toward cooling? Not necessarily, but it certainly does not support the hypothesis of constant warming caused by today's persistent annual increases in atmospheric CO_2 content which are present in Earth's atmosphere 24 hours a day. Nor does the 35 year cooling period from the 1940's to 1970's.

There are some scientists now saying that we can predict the near-term stronger and weaker solar cycles. One current prediction by NASA, 2006, is that the solar sunspots and magnetic activity will be extremely weak around 2022 (see Figure 35). Both of these solar activities have already weakened in 2008. This would allow more cosmic rays to reach Earth's lower atmosphere and possibly create more clouds and cool Earth or it could simply mean less radiation and solar wind coming from the sun. How nice to have a prediction that is for a time less than 15 years away, when the forecaster may still be alive and can be held accountable or congratulated for the prediction. Many, including the IPCC, predicted continued warming 15 to 20 years ago and have been wrong. Some are now changing their views. With satellites now in place to constantly monitor the sun, predictions of future solar influence should improve.

Fig. 35

THE SUN'S CONVEYOR BELT

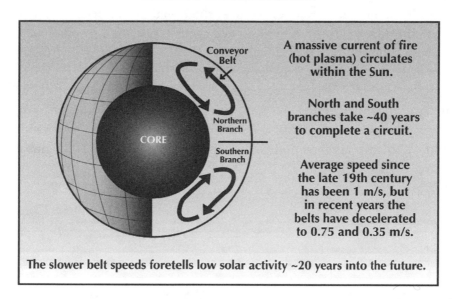

Conveyor Belt

Northern Branch

CORE

Southern Branch

A massive current of fire (hot plasma) circulates within the Sun.

North and South branches take ~40 years to complete a circuit.

Average speed since the late 19th century has been 1 m/s, but in recent years the belts have decelerated to 0.75 and 0.35 m/s.

The slower belt speeds foretells low solar activity ~20 years into the future.

NASA has never seen speeds so slow. "The slowdown we see now should mean the next cycle that will peak around 2022 should be one of the weakest in centuries," per David Hathaway of NASA. Researchers believe the belts control the sunspot cycles. Low sunspot numbers have generally been associated with cooler times on Earth.

Adapted after Science@NASA, Long Range Solar Forecast, May 10, 2006

With confidence I can say the climate will not stay the same. It will continue to change. When you look at any of the examples of historic temperature curves in this book or in anybody else's book or any research article, you won't see any flat lines! The temperature is always either increasing or decreasing. One day, within no more than a few thousand years, Earth will enter another glacial cycle unless researchers can develop the powerful technology required to overcome the tremendous natural forces that have caused Earth's climate changes for the last several million years.

The good news is that with most of these scenarios, the human race can make a lot of adjustments in a few thousand years and avoid suffering the catastrophic consequences depicted in the media.

If you are prone to worry (some seem to enjoy it), go back and look at the many examples of temperature up-spikes that have occurred over very short periods of time such as those that are documented in the Greenland ice cores. Of course, these mainly occurred as Earth was entering, or in the grip of, an ice age. While the up-spikes were usually followed by rapid down-spikes, I think I know which spikes our ancestors preferred (see Figure 32, page 70)!

Stay tuned. There are many unknowns, and problems to be researched and solved. A prime example is found in the ongoing arguments about something as pervasive and important as water vapor and whether it has a net positive or negative feedback impact on global climate.

Chapter Nine

CONCLUSION AND AUTHOR'S OPINIONS

Earth is dynamic, not static; witness volcanoes and earthquakes, continuously shifting continents, sea floor spreading centers, shifting ocean currents, changing sea levels, and ever-evolving orbital alignments - always in a state of change. Because the major climate drivers operate on many different timescales it is easy to see why constant change is to be expected. The changes can and have occurred in both directions and at small and large magnitudes and at fast or slow pace. To see just how frequently the changes have occurred over different time intervals look in Figure 36, page 110. Since there are no flat lines on any timescale, it does support the fact that Earth's climate is dynamic and never in a state of equilibrium.

Even if we do not add one more molecule of greenhouse gas (water vapor, carbon dioxide, methane, nitrous oxide or ozone) to the atmosphere, the climate will still change. And these climate changes could very well send us in a direction that is not necessarily ideal for the future of humanity. Today, this is the reality and there is not a whole lot we can do about it. In the meantime, we should not perturb the natural forces unless it is in a direction we are confident will be beneficial.

Should we continue research on how best to capture some of the CO_2 we create and then store it for possible future use? I have already voted "yes" on this. It is part of the "Don't perturb the system" argument, and it addresses the "Warmer is uncomfortable but colder is catastrophic" reality.

Fig. 36

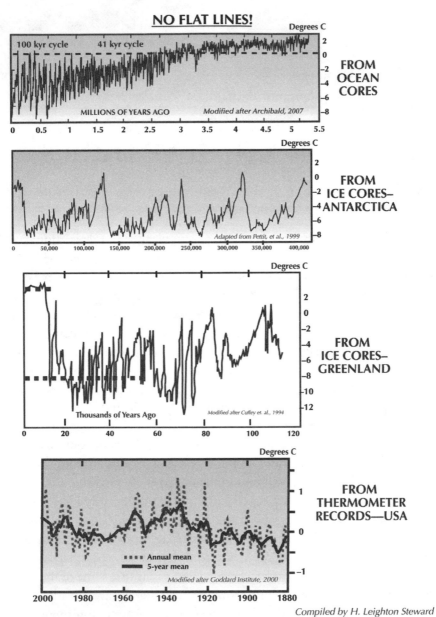

Compiled by H. Leighton Steward

Earth has accumulated a vast supply of fossil fuels such as oil, gas and coal. The use of these concentrated sources of energy has brought much of humanity to a very high standard of living, including better nutrition and health. However, with more than six billion people on Earth, expected to increase to nine billion in just forty years, the supply of this relatively cheap, efficient energy is being deleted at a rapid rate.

I do believe strongly in developing more renewable energy. But I mean developing it at a reasonable pace, not rushing blindly in and upsetting our overall economy and ability to feed, clothe, shelter, and provide medical care for ourselves and much of the rest of the world. Today, five billion of the six and a half billion people who live on Earth are living in somewhat primitive conditions. Electricity, adequate nutrition, and proper medical care are critically needed. I don't want us to harm inadvertently a portion of the world's population by taking money away from these other areas in need of research and assistance by making draconian, mandatory reductions in global CO_2 emissions.

Many people are clamoring to increase the use of our edible grains, which barely support today's population, as an inefficient, expensive fuel source that has little, if any, positive effect on greenhouse gas reduction. Where is the common sense in this? It does make sense to develop new fuels from renewable sources such as cellulose that do not take food from tables or food from the animals.

While I believe the world's people deserve a chance to obtain these life-saving necessities of electricity, clean heating, cooling, and cooking power from today's inexpensive energy supplies, I also believe we should vigorously fund and support the development of more efficient appliances and "engines" of all types and the various mechanisms that use the power from these energy "engines".

Let's get the most from our dollars and brain power. Going immediately to higher-cost, less efficient power will strain the world's economies. While the developed countries can afford some of these added costs, any significant reduction in the use of these most efficient and inexpensive energy sources will only ensure that the people in developing nations will not see significant near-term improvements in their lives and well-being. Let's revisit nuclear energy, which has been proved to be safe, affordable, and non-polluting, if we can simply agree on a suitable storage place for the nuclear waste.

We should support more pure research on how we can best react if natural factors exert a cooling influence on Earth. We may need to "blacken" the surface of Earth to lower Earth's albedo or create more swamps to add more methane to the atmosphere.

What do I believe regarding the issue of global warming? Paleoclimate studies indicate that Earth has warmed somewhat over the last century. These studies also indicate that Earth has cooled over the last 1,000 years relative to the Medieval Warm Period and then started to warm before any significant change in CO_2 content. The climate is in a constant state of change and all of Earth's climate history bears that out. I believe orbital effects, which have been the dominant driver in the past eight out of eight glacial - interglacial cycles, are moving in the direction of pushing the climate of Earth toward another ice age in no more than a few thousand years. CO_2 in the atmosphere, although twice its glacial-period level, remains near its historical low (see Figure 4, page 4) and we have found that it follows, not leads, the temperature changes. We also know that the heat-trapping capability of each additional unit of CO_2 declines rapidly relative to the capability of the previous unit put into the atmosphere. These last two sea-changing or paradigm-changing revelations are causing a large number of climatologists, paleoclimatologists, and scientist in related fields to look at CO_2's effect on climate in a very different light. There may be other such paradigm-changing discoveries that will come from the researcher's laboratories, astronomical observatories, or solar or even extrasolar system space probes.

We need to determine, if possible, at what atmospheric saturation level CO_2 becomes a nonissue relative to further warming and whether that saturation level has already been reached (see Figure 16, page 28). That said, I am not advocating pouring an unlimited amount of greenhouse gases into the atmosphere because the unlimited addition of the gases might affect in a negative way the chemistry of Earth's oceans or some other as yet unidentified environmental impactor. More research is needed in the area of CO_2's effect on ocean chemistry and aquatic life's sensitivity to any such changes.

The near-term movement of global climate into a new cold cycle, short of a huge extraterrestrial impact or a super volcanic explosion, could be the most destructive event for the six and a half billion people on Earth that we can imagine. How near? In our or our children's or grandchildren's lifetimes? I'm not predicting exactly when it will occur, because I simply don't know. I'm just saying I believe it will happen within no more than the next several thousand years.

What else can we do? Now that Earth's library is more interpretable, let's accelerate our research of the past so we can better understand what has previously happened in reaction to changes with these individual and collective climate drivers. Then we will better know what to do if confronted with a situation that could negatively affect us or our descendants. Let us heed the advice from Winston Churchill, who reportedly said, "The farther backward you look, the farther forward you are likely to see." That look back tells me that Earth's climate has never been stable and today's climate changes are normal.

What should we not do? We should not spend "trillions of dollars" on a quick fix for a potential problem that has not been openly debated in a neutral forum by qualified experts. In my opinion, humans are much more likely to vanish from this Earth as a result of a virulent and resistant disease for which we have no cure, such as HIV, than we are from either global warming or global cooling. Let's spend the lion's share of the "trillions of dollars" on improving overall living conditions for humanity, and carve out a smaller but significant portion for continued research on global climate change which is still poorly understood.

We should not move forward based on emotional appeals or feelings that have not stood the test of scientific debate. While many reputable scientists have publicly accepted the earlier consensus that CO_2 is the dominant driver in global warming and are now reluctant to say they might have been wrong, thousands of others have stepped forward and admitted that in light of new scientific evidence, they have changed their opinion. Of course, I may be wrong in downplaying CO_2's future warming influence, but I believe this look back at Earth's earlier history and the associated empirical data support my conclusions. I hope you and many scientists, as well as the legislators that distribute research funding, will take this summary data and consider the implications. Remember some sage advice from an early proponent of common sense; Mark Twain said, "It ain't what you know that gets you into trouble, it's what you know you know for sure that just ain't so." Does this sound familiar? Beware of those who profess to know for sure but will not participate in open debates with scientists or others who have a different view or interpretation of the data currently available.

Discussions of global climate change, particularly the current global warming trend, bring on spirited debates about the causes of the warming in private gatherings but, as I have said, I have been very disappointed in the lack of public debate among top scientists in the field of climatology. If I have written of facts or issues that seem controversial, I hope it is a catalyst for more public discussion. Demand the debates and demand that the moderators of the debates be fair and balanced and not trying to steer the conclusions toward their own pre-disposition to catastrophic forecasts.

Since all climate drivers are on a different timescale relative to their effect on climate change, I have made a summary chart to remind you which ones we need to worry about today, or in the next few decades, and which ones your heirs will not need to be concerned with for millions of years. The chart also lists which drivers we can try to influence and which ones we cannot change (see Figure 37, page 116).

As for the scenarios of the future, there are plenty to tease your imagination. There are people out there, or even reading this book, that may have the genius to imagine all of the interactions at once and to predict what measures we should take to try to counter the ill effects of the negative scenarios. Maybe it's just a "computer whiz" who finally comes up with the model that considers all significant factors and assigns them the proper weighting and spits out the right recommendations or actions we should take or attempt to ensure a better future.

As I said in the Introduction, one of my goals in writing this book was to provide you armchair paleoclimatologists a chance to review some of the summary charts and figures and better understand what is or was happening in all the areas that influence the climate. New or better data are coming in every day. What we do not want to do is move ahead based on faulty data. So, come join us! Go to school. Go back to school. This is no short term exercise. Your children and grandchildren will be the beneficiaries of still better data that will improve on our current observations. Should things change very rapidly, we will have to go with a course of action we can collectively agree on to try to preserve our earthly paradise. Let's base any such actions on science, not emotion or guilt trips that we have poisoned the Earth with CO_2.

Where and how we spend our money is critically important. Steering where this money goes is primarily decided in Washington, with lesser amounts determined at state levels. Very few of our legislators and administration officials have any education in the natural sciences, much less meteorology. The same is true for the media and also of many of the high-profile intellectuals often chosen to be interviewed. Climate change, and particularly today's global-warming decisions, should not be based on ill-informed, emotional appeals.

Better-informed citizens are important because that is where the votes originate. A lot of pressure can be put on lawmakers and officials who oversee the funding for research. Hence this book, which is intended to give a non-scientific audience and politicians and even judges a general, "soft" scientific review of the most salient factors governing global climate change. We all need to be equipped to come to our own, better-informed conclusions rather than merely being slaves to media manipulation or politically or financially motivated information sources.

Finally, think about how the current hysteria over CO_2-induced global catastrophes is affecting our children and grandchildren. They are constantly presented with frightening climate information, often by ill-informed educators in the schools. It has been said that children today are more afraid of global warming than of terrorists, cancer, or vehicular deaths. Open, vigorous debates by climate experts will bring much broader, more balanced information to the public, politicians, and educators, and through them, to our children. Let's hope everybody, including our children, can enjoy this current climate paradise by seeing how it compares to other periods of Earth's history.

I promised to give you a bullet point list of the key factors relative to Earth's climate change:

Fig. 37

HUMANS' ABILITY TO CHANGE THE CLIMATE DRIVERS

CANNOT CHANGE

HOW SOON MIGHT WE FEEL THE CHANGE

Sun's strength --------------------------------- Immediately
Orbital eccentricity _____ Slowly
Earth's tilt _____ Slowly
Earth's wobble _____ Slowly
Ocean currents _____ Decades or centuries
Plate tectonics _____ Millions of years
Location of continents _____ Millions of years
Elevation of land masses _____ Millions of years
Volcanism _____ Immediately
Atmospheric circulation _____ Constantly
Water vapor _____ Constantly
Cosmic rays _____ Immediately

CAN CHANGE

GHG CO$_2$ _____ Decades or centuries
GHG methane _____ Immediately (clathrates)
Chemical weathering (slightly) _____ Very slowly
E.T. impacts (divert) _____ Immediately
Fauna and flora _____ Decade
Albedo _____ Decade

H. Leighton Steward, 2007

GLOBAL CLIMATE SUMMARY

- Climate is continuously changing; there are no flat spots on any temperature curves.

- Constant change is expected since the climate drivers operate at different time scales.

- Earth has been generally warming for over 250 years and the cause is still being argued; the science has not been settled and public debates are needed.

- Technology today enables the reading of Earth's paleoclimate "library" and gives clues as to causes of these earlier climate changes; demand more research.

- Changes in levels of CO_2 have been determined to follow (lag) temperature changes; a cause does not follow an effect.

- CO_2's ability to trap heat declines rapidly, logarithmically, and quickly reaches a point of significantly reduced effect. Many scientists say we are already at the point of insignificance. This, the lag effect, and the negative correlations of temperature and CO_2 from empirical observations in the not too distant past, denies the "CO_2 is the dominant cause of global warming" hypothesis.

- The public's infatuation with CO_2 as the primary cause of climate warming is inappropriate but we should still begin to reduce our rate of input and sequester what is technically and economically reasonable in case there are non-climate negative effects not currently recognized.

- The supply of fossil fuels is limited and we should vigorously pursue improved efficiencies of energy dependent devices and development of reasonable alternative fuels.

- Modelers should be evaluating all the key drivers that affect climate change and researching what mankind might be able to do to temper negative climate changes.

Lastly, reflect on the data presented in *Fire, Ice and Paradise* and then heed another sage quote (see Figure 38). Also included is a discussion on myths that can be found in Chapter 10 that covers my views on many questions you may have pondered.

Fig. 38

IF THE FACTS CHANGE, I'LL CHANGE MY OPINION.

WHAT DO YOU DO, SIR?

John Maynard Keynes

Chapter Ten

POPULAR MYTHS OR GROSS EXAGGERATIONS

Myth: The graph by Mann, et al., the so-called "hockey stick" used in the 2001 Intergovernmental Panel on Climate Change (IPCC) report, showing 1,000 years of stable temperatures until the twentieth century when the temperatures rose rapidly after humans began putting large quantities of CO_2 into the atmosphere, proved that CO_2 has caused global warming.

Fact: Later, the IPCC rejected the validity of the Mann graph, since it left out the data that showed the climate had varied considerably over the past 1,000 years before humans introduced significant amounts of CO_2 into the atmosphere (see Figure 39, page 120). Since CO_2 was shown to have stayed relatively level at 280 *ppm* for these same 1,000 years before the large input of CO_2 by industrialization in the mid-1900s, the corrected graph indicates the opposite, that CO_2 has not been the driver of climate change.

Myth: The close correlation of CO_2 and temperature, as temperature has gone up and down over the last 400,000 years, proves that CO_2 is causing the climate changes (See Figure 12, page 21).

Fact: Since 1999, multiple technical, peer reviewed articles have been available that demonstrate exactly the opposite conclusion; that CO_2 lagged temperature changes as temperature increased or decreased. Temperature changed and then, several hundred years later, CO_2 levels changed. Since a cause does not follow an effect, this indicates that CO_2 is not a primary driver of climate change.

Myth: Sea level likely will rise 20 <u>feet</u> by the end by the century.

Fact: Even the IPCC says the most likely rise will be 17 <u>inches</u> while most climatologists predict a rise of only 7 or 8 inches (about 3 millimeters a year).

Myth: Scientists are unanimous that man-made CO_2 is the dominant cause of global warming.

Fact: Not so. Many, many reputable scientists believe that natural factors overpower the current influence of CO_2 on global warming. Several hundred prominent scientists and/or science professors that have no ties to the petroleum industry have stated publicly that CO_2 is not a significant cause of global warming (Figure 40, page 121 and Figure 41, page 122).

Fig. 39

WHICH GRAPH DO YOU BELIEVE?

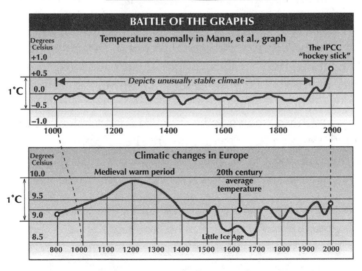

A graph appeared in the Intergovernmental Panel on Climate Change (IPCC), 2001 report (grossly generalized to the top curve above), to "prove" man's impact on global warming (the sudden upward temperature rise), rather than the historically documented graph (lower curve), that shows large climate swings prior to any significant contribution of greenhouse gases by man. IPCC subsequently withdrew Mann's graph because of improper use of data and statistics in its construction.

Modified after McIntyre, et al., 2005

Fig. 40

IS THE SCIENCE SETTLED?

In December, 2007, 100 prominent scientists sent a letter to the Secretary-General of the United Nations and the Bali Climate Conference urging no precipitous action regarding greenhouse gas reduction. The scientists believed the catastrophic forecasts of global warming to be false and that mankind would suffer if greenhouse gases were reduced immediately. The scientists were from 18 different countries and dominantly professors. Thirty-one were trained in meteorology–climatology, 17 were physicists, 16 were geologists and 13 were engineers. At least 91 of the 100 were PhDs.

Some of the more prominent ones included:
- A Global Laureate 500 in meteorology
- President of the Commission for Climatology, World Meteorological Association
- Head of Forecasting Center, Meteorology Institute of Norway
- President of World Federation of Scientists, Switzerland
- Director of Research, Royal Netherlands Meteorological Institute
- Director of the Weather Satellite Service
- President of the European Association of Science Editors
- Director of the Australian National Climate Center
- President of the Center for Study of CO_2 and Global Climate Change
- Director of the Netherlands Institute for Isotopic Geosciences
- Director of the Institute of Antarctic and Southern Ocean Studies

The following week 400 more scientists were listed as having publicly stated that they believe natural, rather than man-made influences, are controlling global climate change.

In June of 2008, a petition was circulated by the Oregon Institute of Science and Medicine that included signatures of 31,072 scientists, 9,021 of whom were PhDs, that said the issue of man-made CO_2 causing global climate change was not settled.

The science is NOT settled. The current global warming mania is going to cause "trillions" to be wasted or misdirected, damaging the United States the most. Public, moderated debates are badly needed.

I am not alone but do have a different, educational approach to the issue of what is causing global climate change. ■

H. Leighton Steward

Fig. 41

AM I ALONE IN THESE VIEWS?
"NO!"

– President of the World Federation of Scientists and Consultant to the Vatican

– President of the Commission for Climatology, World Meteorological Association

– Director of the Paris Institute of Global Physics

– Director of Space Research at the Russian Academy of Science

– Chairman of the Science Committee, 2008 International Geologic Congress

– First woman with a PhD in Meteorology, called one of the most pre-eminent scientists of the last 100 years

– President of the European Association of Science Editors

– A Global Laureate 500 in Meteorology

– UN-IPPC Japanese Scientist in environmental physical chemistry

– NASA Astronaut and Physicist Walter Cunningham of Apollo 7

– Dr. John Theon, Senior NASA Atmospheric Scientist and Supervisor of NASA's James Hansen (man-made catastrophist).

Behind these is a cast of several 10,000s.

AM I ALONE IN THESE VIEWS?
"NO!"

WHAT ARE SOME OF THESE PROMINENT SCIENTISTS NOW SAYING PUBLICLY?

■ "When people come to know the truth, they will feel deceived."

■ "Warming fears are the worst scandal in history."

■ "It is a blatant lie that only a fringe of scientists do not buy man-made warming."

■ "Many scientists are trying to back out quietly without their careers ruined."

■ "CO_2 emissions make absolutely no difference, but it doesn't pay to say so."

■ "Almost none of the recent warming is from CO_2."

■ "Contrary to the IPCC summary report, there is no man-made signature of global warming."

■ "The climate models have been dismal failures."

■ "The climate models are meaningless mush."

■ "I was dismayed the IPCC misrepresented my committee's report."

■ "Following doomsday scenarios can have grave consequences for the Earth's poor."

■ "Hansen is a political activist who spreads fear, even when NASA's own data contradict him."

■ "Hansen embarrassed NASA with catastrophic climate forecasts. The NASA models were useless. Hansen (and others) manipulated the data to get their desired results."

H. Leighton Steward
After Poznan Conference Update
and Solomon's "The Deniers"

Myth: Higher temperatures kill people.

Fact: A half-truth at best. High temperatures can cause deaths but, according to health officials, many more people die each year from cold weather than from hot weather. Just read the news after the strong winter storms.

Myth: Warmer weather will cause more malaria.

Fact: According to Dr. Paul Reiter of the world-renowned Pasteur Institute, this is not true (Reiter, Paul, 2009, personal communication). Malaria was much more prevalent and had already extended into Siberia before the most recent warming.

Myth: Global warming will adversely affect worldwide food production.

Fact: Exactly the opposite. Warmer weather will open much more land to agriculture (at higher latitudes) and the rising CO_2 will stimulate plant growth and the plants will require significantly less water. (Idso, Craig, 2009, personal communication).

Myth: The current rate of warming is unprecedented.

Fact: The short-term temperature swings during the most recent glacial stage had much higher rates of cooling and warming. Also, the rate of temperature rise in the 1920s and 1930s, before people began putting huge amounts of CO_2 into the atmosphere, was higher than the rate of increase over the last 25 years of the 20th Century (see Figure 42, page 124).

Myth: The United States is the largest contributor of man-made CO_2.

Fact: China, which has no CO_2 restrictions per the Kyoto protocol, has recently exceeded the United States in CO_2 output and the gap is growing very rapidly because China is opening a new coal-fired power plant every WEEK and its production of automobiles is growing at a much more rapid rate than that occurring in the United States.

Fig. 42

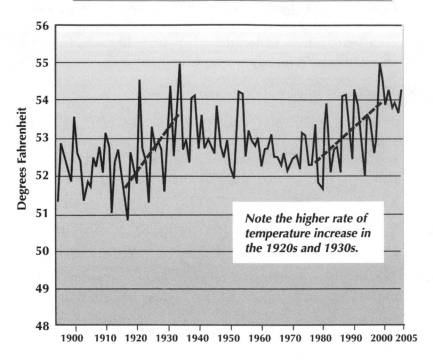

UNITED STATES AVERAGE ANNUAL TEMPERATURE

Note the higher rate of temperature increase in the 1920s and 1930s.

Our climate roller coaster continues.

SOURCE: National Climatic Data Center
Modified after Christopher C. Horner, 2007

Myth: Storms are now much more frequent and intense because of global warming.

Fact: According to the National Hurricane Center, storms are no more intense or frequent worldwide than they have been since 1850. Temperatures were high in the 1920s and 1930s when there was much less CO_2 in the atmosphere. Constant 24/7 media coverage of every significant storm worldwide just makes it seem that way (See Figure 43). Insist on the facts, not just what some individuals or reporter say to support their cause.

Fig. 43

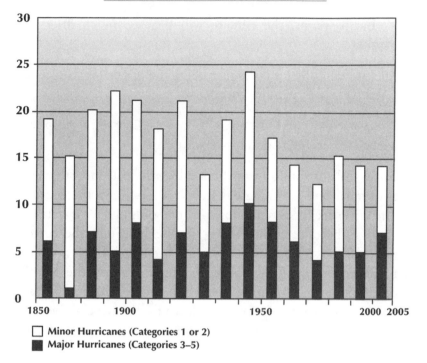

U.S. HURRICANE STRIKES BY DECADE

☐ Minor Hurricanes (Categories 1 or 2)
■ Major Hurricanes (Categories 3–5)

Hurricane numbers and intensities have not increased with more CO_2 or warming.

SOURCE: National Climatic Data Center
Modified after Christopher C. Horner, 2007

Myth: Large chunks of ice breaking off glaciers (calving) around the world are proof of the catastrophe being caused by global warming.

Fact: Glaciers constantly calve ice into the sea even when the glaciers are growing. Photographs of such activity are no proof of any catastrophe occurring.

Myth: Human activities are causing unprecedented melting of Earth's glaciers and accompanying sea level rise.

Fact: Ask yourself, why did so many more glaciers melt and sea level rise 18 to 20 feet higher than today during the last interglacial period, 120,000 years ago, when modern man was not around to introduce any CO_2 into the atmosphere? Ponder the facts.

Myth: As Earth warms, the climate will become much drier and windier.

Fact: All studies of the ice cores prove just the opposite. The colder times were both windier and drier.

Myth: For every additional unit of CO_2 that we add to the atmosphere, we will see an equal amount of warming.

Fact: False. Calculations by many reputable scientists show that each additional unit of CO_2 will trap less heat than the previous unit. The decline is rapid, logarithmic, causing CO_2's warming effect to become minimized very quickly. Since the current pace of the temperature rise has been less than modelers predicted from the ongoing CO_2 rise, it supports the premise of CO_2's diminishing role in causing climate change. In fact, Earth's temperature has trended down for the last eight years as worldwide CO_2 levels have continued to rise (Figure 20, page 33).

Myth: Polar bears will likely go extinct if this warm period continues through the 21st Century.

Fact: A jawbone of a polar bear has been found that is 120,000 years old, a time during the previous interglacial when temperatures were five degrees Celsius warmer and sea level 19 feet higher than today. They adapted then, why not now?

Myth: Adding more CO_2 into the atmosphere will cause the ocean to become more acidic.

Fact: The ocean is not acidic, it is alkaline. More atmospheric CO_2 might cause it to become less alkaline but not acidic. If Earth continues to warm, there should be a net release of CO_2 from the oceans.

Myth: It will take only a little more CO_2 to trigger a runaway global warming.

Fact: Earth has never experienced a runaway warming, even when CO_2 levels were 5,000 to 7,000 ppm! Ironically, for the anthropogenic global warmers, this indicates that some of the feedbacks must be negative as Earth warms and becomes wetter and cloudier, which can promote cooling.

Myth: CO_2 is a pollutant.

Fact: CO_2 is a great airborne fertilizer which, as its concentrations rise, causes additional plant growth and causes plants to need less water. Without CO_2, there would be no life (food) on Earth. The 100 ppm of CO_2 added to the atmosphere since the start of the industrial revolution has caused an increase in worldwide plant growth of 12 to 15 percent.

RECOMMENDED READINGS

Summarized below are a few books that address one phase or another of global climate change. All are relatively informative, and several are very well written. One textbook is included for those of you who want to know what you would learn if you took an intensive college-level course in this subject or, better yet, pursued a major in this field. These books are referenced in the bibliography even though they may not be specifically referenced in the text. How, for instance, could a researcher ever figure out who first coined the term "driver" or "indicator", or spoke of CO_2 being an important greenhouse gas or that climate changes have affected past civilizations? If I knew, I would certainly reference them. Two important factors are missing from several of these books that mention global warming. One is that the CO_2 levels follow temperature changes, and the other is that CO_2 quickly loses its ability to trap more heat, facts that may not have been known or recognized at the time some of the manuscripts were in production.

A Brain for All Seasons by William H. Calvin, M.D. (2002). Dr. Calvin, a professor of psychiatry, writes a great summation on global climate change and how the constant stress caused by the changes resulted in human brains either evolving to cope with the change or becoming another dead-end species or line of individuals.

The Eternal Frontier by Tim Flannery (2001). This is a paleohistory of North America from plants to animals to early humans. It is very well written, and it should cause you to look at this continent in a way that will make you want to go sit on a Wyoming mountaintop and dream of what North America looked like all those years prior to our ancestor's recent arrival.

The Two Mile Time Machine by Professor of Geosciences, Richard B. Alley (2000). Dr. Alley was directly involved in coring and interpreting a two mile long ice core taken from the Greenland ice sheet. His analyses helped alert the world that many global climate changes have occurred in periods as short as a few years to a few decades. Alley's book was probably in production before it became well known that atmospheric CO_2 changes followed temperature changes.

The Ice Chronicles by Paul A. Mayewski and Frank White (2002). Paul Mayewski was the chief scientist on the expedition that drilled the two-mile ice core mentioned above. Frank White is a writer. Mayewski believes global warming is caused by humans but advocates a fairly balanced approach to greenhouse gas reduction accompanied by more research.

Geological Perspectives of Global Climate Change by Lee C. Gerhard, et al. (2001). This is a compendium of chapters by various scientists that addresses multiple causes and effects of climate change and demonstrates that the climate has changed continuously before any influence by humans. It also contains a chapter by Sherwood Idso that summarizes 20 years of real world research designed to determine Earth's sensitivity to changes in the levels of greenhouse gases.

Out of Thin Air by Peter D. Ward (2006). Ward presents a very different viewpoint on the evolution of plant and animal life in that he postulates that the changing levels of atmospheric oxygen were the driving mechanism for evolutionary change and that those species that adapted well, such as the dinosaurs, became the dominant species of their time.

Earth's Climate, Past and Future by William F. Ruddiman (2000). This is a college-level textbook and the best written and illustrated of any of the several I have read. Ruddiman's coverage of global climate is broad, well documented, and quite readable and understandable for someone with a modest background in science. It was, however, apparently going into print before Fischer's announcement that temperature changes precede the changes in levels of CO_2. Otherwise he would have mentioned it and its shocking implications.

Unstoppable Global Warming, Every 1,500 Years by S. Fred Singer and Denis Avery (2007). This is a very good accounting of factors, fears, and myths regarding man-made global warming by a prominent climate physicist and professor (Singer), aided by a well-known economist and writer (Avery). They document the natural cycle of warmth that has occurred every 1,500 years due, they say, to variations within the Sun.

The Deniers, by Lawrence Soloman (2008). Soloman, an environmental activist and columnist, purposely set out to interview many of the most prominent scientists and researchers in the world who have become skeptics of man-made global warming. He documents the essence of the interviews and offers his opinions regarding the qualifications of those interviewed. He summarizes that "straight down the line"; the deniers have better scientific credentials that the man-made global warming advocates. This is a "must read" for anyone who thinks the science on global warming is settled. Even a panel of the National Science Foundation found that the science is not settled.

Cool It, by Bjorn Lomborg (2008). Lomborg, a world class economist, uses both facts and common sense to recommend what governments should do relative to the issue of global warming. Another "must read" to help you set your priorities when considering the funding of the research or action to be taken on many of the major challenges we face today.

ACKNOWLEDGEMENTS

An acknowledgement is often taken as an indication that the people who have encouraged or assisted the author tacitly agree with the conclusions expressed in the author's book. The conclusions reached in this book are strictly my own. I have no problem with anyone who has come to a different interpretation of any of my conclusions, and I encourage you, the reader, to utilize the scientific data presented in this book, plus any other scientific data you may find, to arrive at your own conclusions. But do not take the statements as meaning I have unexpressed concerns about my conclusions or recommendations, because until new scientific data come in to change my mind, these are my conclusions and recommendations and I'm sticking with them!

The first acknowledgment is to Dr. Crayton Yapp, professor of geological sciences at Southern Methodist University (SMU), who is an expert in the use of stable isotopes to determine the paleoclimate history of Earth. Crayton's telling me of CO_2 levels in the atmosphere hundreds of millions of years ago that were ten or more times higher than in the atmosphere today got me interested in this subject. After a cursory look at how the temperature of Earth, as well as the concentrations of greenhouse gases, have moved up and down over the eons, I thought it would be useful to try to summarize those facts for the general, nonscientific public as well as for our policymakers, few of whom have any science background. This leaves most people at the mercy of the cryptic summaries provided by the media or by constituents who may have their own interests in mind.

After they saw a crude outline of my proposed book (no conclusions yet), I received encouragement, questions, and advice from Dr. Jim Brooks, one of my geology professors, and Dr. Louis Jacobs. These two very fine gentlemen are vice chairman and president, respectively, of the Institute for the Study of Earth and Man at SMU, an organization that has a mission of bringing mankind and the environment together through the better understanding of geology, archeology and anthropology. I received further assistance from Dr. Bob Gregory, chairman of the department of geological sciences at SMU, Professor Crayton Yapp and Associate Professors Bonnie Jacobs and Neil Tabor. Spirited debates ensued! As I said, the conclusions in this book are mine.

Diana Vineyard, a Master of Science candidate about ready for prime time, helped me find many of the scientific articles that were important to my research.

On the domestic front, my wife of several decades, Lynda, who was a fashion design major, read an early draft and bounded in smiling. I took that to mean that a non-scientist might be able to grasp the essence of global climate change. Lynda named my first book (Sugar Busters ®), which with its offsprings has sold more than four million copies. I know, I know, this book has no such chance because it will not help you lose weight and improve your blood chemistry, but it may also change the way you and your descendants live in the future. Lynda has had to put up with my stacks and boxes of research papers piled on her dining room table, kitchen table, and bar for more than two years. Her early enthusiasm for my writing of *Fire, Ice and Paradise* did not waiver even when she realized she had lost me, not for three or four months, but for more than two years.

I received encouragement from several of my old geology friends, Dr. Gene Shinn, Gene Ames, Joel Wilkinson, Jim Nielson, Jim Gibbs, Dr. Pete Rose, Bob Walpole and Dave Haglund, who saw early versions of the text or the illustrations. Dr. Gene Shinn and Gene Ames were particularly helpful by furnishing me scientific reports I might otherwise have overlooked. I received some very good information from Drs. Lee Gerhard and Ray Thomasson. Lee has been a lead author and lecturer on global climate change and Ray probably will write a more technical book on this extremely complex subject. None of these fine folks believes that the scientific community has come to a consensus on what is the major cause of global warming or global climate change.

Jan Woods-Krier professionalized my illustrations and patiently put up with my many tweaks and changes. Maurine Maloy gave me prompt turnaround on the early and final drafts of my manuscript. My editor for *Sugar Busters!* Maureen O'Neal, who had learned to read my scribbling previously, gave me both editorial and common sense advice.

In writing a non-fiction book that includes some technical illustrations, I discovered an "eternity" factor. It takes an eternity to get permissions to use some of the illustrations. I would still be waiting if not for the advice and help of another fine lady, Mary Kay Grosvald, librarian for the AAPG in Tulsa, Oklahoma.

Lastly, I thank everyone who saw the need and then encouraged me to write a book on the soft science of this subject written at the layman, legislative and administrative level.

Thank you all!

GLOSSARY

Aerosols – very small particles or droplets suspended in the air. They are conducive to forming water droplets and clouds.

Albedo – the percentage of incoming radiation from the sun that is reflected away from the Earth's surface or atmosphere.

Amphibian – a form of animal life that spends part of the time in the water and the other part on land.

Anthropogenic – caused by human activities.

AGW – anthropogenic (human-caused) global warming.

Aquatic – growing or living in or upon the water.

Axial precession – the wobbling of Earth's rotation (spinning) that causes Earth's axis of rotation (aligned from North to South Pole) to point in different directions over cycles of 19,000 to 23,000 years.

Carbon dioxide – a greenhouse gas made of one part carbon and two parts oxygen (CO_2).

Carbonate rocks – rocks like limestones made primarily of one part calcium, one part carbon and three parts oxygen ($CaCO_3$).

Celsius – a temperature scale on which water at sea level freezes at zero degrees and boils at 100 degrees.

Chemical weathering – alteration or dissolving of minerals to a different form when exposed to contact with water and the atmosphere.

Clathrate – a form of matter in which a substance such as methane resides in a frozen state.

Comet – a frozen mass of dust and gas orbiting the sun. Comets occasionally collide with Earth or Earth's atmosphere.

Continental crust – the outermost layer of Earth's surface that averages approximately 20 miles (35 kilometers) thick beneath the continents.

Continental drift – the lateral movements of rigid (lithospheric) plates that lie above the molten mantle layer within Earth that occurs at approximately 60 miles (100 kilometers) below Earth's surface.

Convection cells – motions within the molten layers of Earth where hotter molten areas rise and cooler areas sink, forming circular loops that cause the rigid plate above to move or drift.

Cosmic rays – high-energy particles from outer space that bombard Earth and Earth's atmosphere and produce secondary microparticles in Earth's lower atmosphere, some of which are believed to help form low level clouds that cool Earth.

Crustal plates – the outer layers of Earth that average about 3.5 miles (6 kilometers) thick beneath the oceans and 20 miles (35 kilometers) thick beneath the continents.

Drivers – used herein as a cause or reinforcement of climate change, including global warming.

El Niño – a climatic pattern that develops in the eastern tropical Pacific every two to seven years that is marked by warmer-than-average surface waters that produce a significant regional, if not global, climate effect.

Elliptical orbit – not round but more oval in shape and not centered equally around the object being orbited (in this case the sun).

Evaporites – rocks or minerals that form by precipitation of crystals from water evaporating in restricted basins in arid climates.

Faculae – bright rings or halos that surround the dark sunspots on the surface of the sun. While the sunspots themselves produce a cooler-than-average surface, the extra energy produced by the faculae more than makes up for the cooler sunspots.

Fahrenheit – a temperature scale on which water freezes at 32 degrees and boils at 212 degrees.

Fauna – the animal life of Earth.

Feedback – a process in Earth's climate system that amplifies or moderates changes occurring in the climate.

Flora – the plant life of Earth.

Gondwana – an early, pre 300 *mya* supercontinent located in the Southern Hemisphere.

Greenhouse effect – the warming of Earth's surface or atmosphere by gases such as water vapor, CO_2, or methane temporarily absorbing some of the long-wave, infrared heat radiated from Earth's surface that helps keep Earth's near-surface temperature above freezing.

Greenhouse gas – an atmospheric gas that temporarily absorbs and radiates infrared energy (heat) and helps keep Earth's near-surface temperature above freezing.

Half-life – the time it takes for one half of the number of atoms of a radioactive isotope to decay.

Hothouse – an extended period of time in which Earth has been, or will be, unusually warm.

Hydrocarbon – an organic compound containing only hydrogen and carbon. Three of the currently most efficient and commonly used energy sources are the hydrocarbons oil, natural gas, and coal.

Icehouse – an extended period of time in which Earth has been or will be unusually cold, e.g. a glacial period.

Indicators – used herein to describe things that give us clues or direct measurements of the older (paleo) climate's temperature and atmospheric composition or energy.

Infrared – radiation made up of non-visible "light" of longer wavelengths. The heat energy radiated from Earth's surface is infrared, non-visible radiation.

Insolation – the amount of the sun's energy (radiation) arriving at the top of Earth's atmosphere.

Ion - an electrically charged particle or atom.

Ionize – to change or be changed into ions, some of which may help in the formation of water droplets that form rain or clouds.

Irradiance – the amount of light or other radiant energy striking a specific surface area.

Isotope – the form of a chemical element that has the same number of protons but a different number of neutrons in the nucleus, which gives the element a different atomic weight.

La Niña – a climate pattern that develops in the eastern tropical Pacific that is marked by cooler-than-average surface waters that produce a significant regional, if not global, climate effect.

Laurasia – a very large continent that formed in the Northern Hemisphere before approximately 300 *mya*.

Lithosphere - the outer, solid portion of the Earth, the crust of the Earth.

Little Ice Age – a period of time from approximately the 1300s to the 1800s, when Earth was colder than the immediately preceding or succeeding periods. The coldest part of the Little Ice Age was coincident with a period of very low sunspot activity that lasted from 1645 A.D. to 1715 A.D.

Luminosity – brightness or the giving off of light.

Maunder Minimum – a time of very low sunspot activity that lasted from 1645 to 1715 A.D. and coincided with (if not caused) the coldest part of the Little Ice Age.

Mechanical weathering – the physical breaking up, exposing, washing away, or blowing away of rocks or sediments.

Medieval Warm Period or Climate Optimum – a period around 1100 to 1300 A.D. when the Northern Hemisphere was one or two degrees Celsius warmer than today.

Meteor – a solid body of rock or metal from outer space that is orbiting or traveling within the solar system.

Meteorite – a solid body of rock or metal that comes from outer space or the outer solar system and strikes Earth's surface; the surviving pieces are called meteorites.

Mid-ocean spreading centers – the place or elongated ridge where two of Earth's plates are moving apart, with the ensuing void being constantly filled with molten rock from deeper within Earth.

Millennium – one thousand years in length, or describing a major period that contains 1,000 years, such as 1,000 A.D. to 2,000 A.D.

Negative feedback – a process that moderates or counters the original perturbation of Earth's climate.

Nitrous oxide – a colorless, nonflammable greenhouse gas made up of two parts nitrogen and one part oxygen (N_2O).

Oceanic crust – the outermost layer of Earth's surface, lying beneath the oceans, that averages approximately 3.5 miles (6 kilometers) thick.

Outcrop – the exposures at Earth's surface where layers or deposits of rock or sediment can be seen and accessed.

Ozone – a triple molecule of oxygen (O_3) formed by the collision of cosmic particles with ordinary oxygen. Ozone is a weak greenhouse gas that also helps block ultraviolet radiation which can cause cancer.

Pangaea – the supercontinent that was formed when all of Earth's continents coalesced into one giant continent, after about 300 *mya*.

Particulates – very small particles of various origins that can help form water droplets, which can precipitate or form clouds.

Photosynthesis – the process by which plants absorb sunlight and use it to convert water and CO_2 to carbohydrates (plant tissue) and release oxygen in the process.

Plate tectonics – effects caused by the lateral motions of the rigid, outermost layers of Earth. These individual or collective plates collide and cause mountain building, slide past each other, or tear apart (spread).

Pollen – the soft fertilizing elements of a plant, the more rigid walls of which can sometimes survive burial and transformation into solid rock.

Positive feedback – a process that adds to the original perturbation of Earth's climate.

Precession – an effect of a spinning body such as a top (or the Earth) when the torque changes the position of its axis of rotation, generally changing the axis in the shape of a circular cone as it returns to its original position.

Preindustrial – refers herein to the time in the later part of the 1800s before the CO_2 content began to rise significantly due to man's accelerated industrial activities.

Radiation – ultraviolet and visible radiation sent from the sun to Earth.

Roman Warm Period – a period of above average global temperatures that peaked around 2,000 years ago which saw the Alpine glaciers retreat nearly 1,000 feet higher than at present.

Sea floor spreading centers – the same as mid-ocean spreading center.

Sequester – store or put away.

Silicious rocks – rocks that contain one part silica and two parts oxygen (SiO_2). These rocks, when chemically weathered, react and take CO_2 from the atmosphere.

Snowball Earth – a time or times between approximately 850 and 650 *mya* when some glacial evidence is found on all continents, even those near the equator, suggesting most, if not all of the surface of Earth may have been frozen.

Solar system – the sun and all of the planets or smaller bodies that orbit the sun.

Stratosphere – the atmosphere that extends from the top of the troposphere, about 6 miles, up to about 30 miles above Earth's surface.

Subduction zone – the area where two of Earth's rigid plates collide and one plate gets shoved under, or subducted, beneath the other.

Sunspot cycle – a natural 11-year cycle of a varying number of visible sunspots on the surface of the sun. The cycles can and do vary in length and intensity.

Sunspot cycle length – the actual length of a particular sunspot cycle which normally is 11 years.

Tilt axis – the angle between Earth's equatorial plane and the plane of Earth's orbit around the sun. The angle changes from about 21.8 to 24.4 degrees over a 41,000-year time period.

Troposphere – the lower part of Earth's atmosphere that extends up to as much as 35,000 feet (7 miles or 10 kilometers) above Earth's surface.

Ultraviolet light – visible, short-wavelength light. This is the type of light that comes from the sun.

Upwelling – the rising of deep, cooler, and generally nutrient-rich waters to the ocean's surface.

Water vapor – water that is in a gaseous state in the atmosphere. When the water vapor condenses into droplets, it becomes either clouds, rain, or ice crystals.

BIBLIOGRAPHY

Abbott, Dallas, 2006, *NY Times, Science Times*, Nov. 14, 2006.

Abbott, Dallas, 2008, professor and adjunct research scientist at Lamont-Doherty Earth Observatory at Columbia University, Palisades, New York, personal communication.

Alley, R.B., 2000, *The Two Mile Time Machine*, Princeton, New Jersey, Princeton University Press.

Ambrose, Stanley H., 1998, "Late Pleistocene Human Population Bottlenecks", *J. of Human Evolution* 34 (6), p 623-51.

Anderson & Borns, 1997, *AAPG Studies in Geology #47*, AAPG, Tulsa, Ok., p 195.

Archibald, David, 2007, Lavorsier Conference Presentation, Melbourne, Australia, http://www.lavorsier.com.au/papers/Conf2007/Hammer2007.pdf.

Berner, R. A., 2006, "Geocarb-sulf: A combined Model for Phaneroazoic O_2 and CO_2", *Geochem. Cosmochim. Acta* 70: 5653-64.

Berner, R.A., 2004, *The Phanerozoic Carbon Cycle: CO_2 and O_2*, New York: Oxford University Press.

Berner, R. A., 1994, "Geocarb III, A revised Model of CO_2", *Am. J of Science*, v. 291: p 56-91.

Bluemle, J. P., et al., 2001, "Rate & Magnitude of Past Global Climate Changes", *AAPG Studies in Geol. #47*: Tulsa, Ok, p 193-212.

Bond, Gerald, et al., 2001, "Persistent Solar Influence", *Science*, v. 294: p 2130-36.

Bond, Gerald, et al., 1997, "A Persuasive Millennial Cycle", *Science*, v. 278: p 1257-66.

Broecker, W. S., 1991, "The Great Ocean Conveyor", *Oceanography*, v. 4: p 79-89.

Broecker, W. S., 1992, "Thermohaline Circulation", *Science*, v. 278: p 1582-88.

Broecker, W. S., 2001, "Was The Medieval Warm Period Global?", *Science*, v. 291: p 1497-99.

Callion, N., et al., 2003, "Timing of Atmospheric CO_2 and Antarctic Temperature Changes Across Termination III", *Science*, v. 299: p 1728-31.

Calvin, W. H., 2002, *A Brain for All Seasons*, Chicago: University of Chicago Press.

Crowley, T. J. and North, G. R., 1991, *Paleoclimatology*, New York: Oxford University Press.

Cuffey, K.M., et al., 1994, "Calibration of the ^{18}O Isotopic Paleothermometer", *J of Glacialology*, v. 40: p 341-49.

Daly, S.I., 2000, "The Hockey Stick": www.john-daly.com/hockey/hockey.htm.

Dansgaard, W., et al., 1993, "Evidence for Instability of Past Climate", *Nature*, v. 364: p 218-20.

Fans, S. M., et al., 1998, "A Large North America Carbon Sink", *Science*, v. 282: p 442-446.

Fischer, H. M., et al., 1999, "Ice Core Records of Atmospheric CO_2", *Science*, v. 283: p 1712-14.

Flannery, Tim, 2001, *The Eternal Frontier*, New York: Atlantic Monthly Press.

Frakes, L.A., 2005, *Climate Modes of the Phanerozoic*, Cambridge, U.K.: Cambridge University Press.

Friis – Christensen, E. & Lassen, K., 1991, "Length of the Solar Cycle, An Indicator of Solar Activity", *Science*, v. 254: p 698-700.

Gerhard, Lee C., 2001, "Climate Change, Conflict of Observational Science, Theory, and Politics", *AAPG Bulletin*: v. 88, p 1211-20.

Gerhard, Lee C., et al., 2001, "Geological Perspectives of Global Climate Change", *AAPG Studies in Geology #47*: AAPG, Tulsa, Ok.

Gerhard, Lee C., 2007, *Search and Discovery*, www.searchanddiscovery.net//documents/2007/07005gerhard/index.htm.

Gerhard, Lee C., 2007, former principal geologist of the Kansas Geological Survey and State Geologist, personal communication.

Gordon, W. A., 1975, *Journal of Geology*, v. 83: p 671-84.

Gore, Al, 2006, *An Inconvenient Truth*, New York, Rodale Books.

Gradstein, F.M., et al., 2004, *A Geologic Time Scale*, Cambridge, U.K.: Cambridge University Press.

Gregory, Robert, 2007, Chmn. Dept. of Geological Sciences, Southern Methodist University: personal communication.

Guslakov, V. K., 2006, "Ancient Crash, Epic Wave", *New York Times, Science Times*: Nov. 14, 2006.

Hallam A., 1984, "Pre-Quaternary Sea Level Changes", *Earth Planetary Science* 12: 205-43.

Hieb, M. 2003, "Global warming: A Closer Look at the Numbers", in Hieb, M. and Hieb, H., 2003, "Global Warming, A Chilling Perspective":   www.clearlight.com/~mhieb/WVFossils/Ice_Ages.html.

Horner, Christopher, 2007, *The Politically Incorrect Guide to Global Warming*, Washington, D.C.: Regnery Publishing, 350 pgs and personal communication.

Hoyt, D. V., 2006, Personal communication to Gerhard, Lee C.

Hoyt, D. V. and Schatlen, K. H., 1997, *The Role of the Sun in Climate Change*, New York: Oxford Univ. Press, 279 p.

Intergovernmental Panel on Climate Change (IPCC), reports from 1990, 1995, 2001 and 2007, New York: Cambridge University Press.

Idso, Craig, 2009, Chairman, Center for the Study of Carbon Dioxide and Global Change, Tempe, Arizona, Personal communication.

Idso, Keith & Craig, 2004, "The Canary in the Coal Mine", *Science*, v. 7: no. 10.

Idso, Sherwood B., 2001, "Carbon Dioxide Induced Global Warming", *AAPG Studies In Geology #47*: p 317-36.

Jacobs, Bonnie, 2007, Assoc. Prof. Dept. of Geol. Sciences, Southern Methodist University: Personal communication.

Jacobs, Lewis, 2007, President, Institute for Study of Earth and Man, Southern Methodist University: Personal communication.

Keeling, C. D., and Whorf, T. P., 1996, "Atmospheric CO_2 Records from Sites in the SIO Air Sampling Network", Oak Ridge National Laboratory, Oak Ridge, Tennessee, USA.

Keppler, F. and Rockmann, T., 2007, "Methane, Plants & Climate Change", *Scientific American*: Feb. 2007, p 52-57.

Khilyuk, L. F. and Chilingar, G. V., 2003, "Global Warming: Are We Confusing Cause and Effect?", *Energy*; v. 25: p 357-370.

Khilyuk, L.F. and Chilingar, G. V., 2004, "Global Warming and Long Term Climatic Changes: A Progress Report", *Environmental Geology*, v. 46: p 970-79.

Lamb, H. H. 1995, *Climate History & The Modern World*, 2nd Edition, London, U.K.: Routledge, 443 p.

Lean, J., 2004, "Solar Irradiance", NOAA/NGDC.

Linden, Eugene, 2006, *The Winds of Change*, New York: Simon & Schuster.

Lindzen, Richard, 2009, International Conference on Climate Change, March 8-10, 2009, New York City, Personal communication.

Lomborg, Bjorn, 2008, *Cool It*, New York, Vintage Books.

Mackenzie, Fred T., 1998, *Our Changing Planet*, 2nd Edition, Upper Saddle River, N.J.: Prentice Hall.

Mann, M.E. et al., 1999, "Northern Hemisphere Temperatures During the Past Millennium", *Geophysical Research Letters*, v. 26: p. 759-762.

Mayewski, P. A. and White, F., 2002, *The Ice Chronicles*, Hanover, N.H.: University Press of New England.

McIntyre, Stephen and McKitrick, Ross, 2003, "Correction to the Mann, et al., Proxy Data Base and Northern Hemisphere Average Temperature Series", *Energy and Environment*, v. 14, No. 6: p 751-771.

Merritts, D., et al., 1998, *Environmental Geology*, New York: W. H. Freeman and Company.

Milloy, Steve, 2006, JunkScience.com, © April 14, 2006, personal communication.

Monnin, et al., 2001, "Atmospheric CO_2 Concentrations Over the Last Glacial Termination", *Science*, v. 291, No. 5: p. 112-14.

Montanez, I. P., et al., 2007, "CO_2 Forced Climate & Vegetation Instability During the Late Paleozoic Deglaciation", *Science*, v. 315: p. 87-91.

Mudlesee, M., et al., 2001, "The Phase Relations: A More Atmospheric CO_2, Quaternary", *Science*, Rev. v. 20: p. 583-89.

NASA, 2006, Science @ NASA, http://science.nasa.gov.headlines/y2006/10may_longrange.html?list3134.

New York Times, Science Times, 2006, "Ancient Crash, Epic Wave", Nov. 14, 2006.

Pekarek, Alfred, 2001, "Solar Forcing of Earth's Climate", in Gerhard, Lee C., et al., 2001, "Geological Perspectives of Global Climate Change", *AAPG Studies in Geology #47*: p. 19-34.

Petit, J. R., et al., 1999, "Climate & Atmospheric History of the Past 420,000 Years", *Nature*, v. 39: June 3, 1999, p 429-35.

Rahmstorf, Stefan, 2006, *Climate Change Fact Sheet*, www.pik-potsdam.de/~stefan.

Reiter, Paul, 2000, "From Shakespeare to DeFoe: Malaria in England in the Little Ice Age", *Emerging Infectious Diseases* 6: p 1-11.

Reuters, 2007, "Study: Killer Hurricanes Thrived in Cooler Seas".

Robinson, Arthur B., et al., 1998, "Environmental Effect of Increased Atmospheric Carbon Dioxide: Petition Project", La Jolla, CA.

Robinson, Arthur B., et al., 2007, J. American Physician and Surgeons, v.12, 79-90.

Royer, D. L., et al., 2004, "CO_2 As A Primary Driver of Phanerozoic Climate", *GSA Today*, v. 14: p. 4-10.

Royer, D. L., 2006, "CO_2 Forced Climate Thresholds during the Phanerozoic", *Geochim. Acta* 70: 5665-5674.

Ruddiman, William F., 2000, *Earth's Climate, Past & Future*, New York: W. H. Freeman & Co.

Scotese, C. R., 2001, "Atlas of Earth History", v.1, *Paleogeography*, Arlington, Texas: PALEOMAP Project (also http://www/scotese.com/earth.htm).

Segalstad, Tom, 2009, International Conference on Climate Change, March 8-10, 2009, New York City, Personal communication.

Shaviv, N. J. & Veizer, J., 2003, "Celestial Drivers of Phanerozoic Climate?" *GSA Today*, v. 13: no. 7.

Shaviv, N. J., 2003, "Towards a Solution to the Faint Young Sun Paradox", *Journal of Geophysical Research*, v. 108: p 1437.

Shinn, E. A., 2001, "Coral Reefs & Shoreline Dipsticks", in Gerhard, L. C., et al., *AAPG Studies in Geology #47*, 2001: p. 250-64.

Shinn, E. A., 2009, Retired U.S. Geological Survey Pioneer researcher in coral reef ecosystems, Univeristy of Southern Florida.

Siegenthaler, U., et al., 2005, "Stable Carbon Cycle Climate Relationship", *Science*, v. 310, p. 1313-17.

Singer, Fred S. & Avery, D. T., 2007, *Unstoppable Global Warming*, Lanham, Md.: Rowman, & Littlefield.

Skinner, Brian J. & Porter, Stephen C., 1995, *The Blue Planet*, New York: John Wiley & Sons, Inc.

Smith, Robert & Siegel, L. J., 2000, *Windows Into The Earth*, New York, Oxford University Press.

Solanki, S. K., et al., 2004, "Unusual Activity of the Sun During Recent Decades Compared to the Previous 11,000 Years", *Nature*, v. 431: p 1084-87.

Soon, W. S. et al., 2001, "Modeling Climatic Effects of Anthropogenic Carbon Dioxide Emissions: Unknowns and Uncertainties", *Climate Research*, v. 18: p 259-275.

Soon, W., et al., 2003, "Reconstructing Climate & Environmental Changes of the Past 1000 Years: a Reappraisal", *Energy & Environment*, v. 14: p 233-96.

Svensmark, Henrik, 2007, "Cosmoclimatology: A New Theory Emerges", *A & G*, v. 48, p.18-24.

Svensmark, H. & Calder, N., 2007, *The Chilling Stars*, Chambridge, U.K., Icon Books, Ltd.

Tabor, Neil, 2007, Assc. Prof. Department of Geological Sciences, Southern Methodist University, personal communication.

Thomasson, M. Ray, 2007, Thomasson & Associates, Denver, Colorado, personal communication.

Usoskin, Schussler, Solanki and Murssaula, 2006, "Solar Activity, Cosmic Rays and Earth's Temperature: a Millennium-Scale Comparison", *J. of Geophysical Res.*, v. 110: A 10102.

Vail, P. R., et al., 1977, "Seismic Stratigraphy and Global Changes of Sea Level", *AAPG Memoir* 26: p 83-87.

Vakulenko, N. V., et al., 2004, "Evidence for the Leading Role of Temperature", *Dolk. Russian Academy of Science, Earth Sciences*, v. 397: p 663-67.

Von Frese, Ralph, 2006, "Big Bang in Antarctica – Killer Crater Found Under Ice", *Ohio State Research News*, Columbus, Ohio.

Ward, Peter Douglas, 2006, *Out of Thin Air*, Washington, D.C., Joseph Henry Press.

Winchester, Simon, 2001, *The Map That Changed the World*, New York: Harper Collins.

Yapp, Crayton, 2004, "$Fe(CO_3)OH$ in Geothite from a Mid-latitude North American Oxisol", *Geochemica & Cosmochemica*, v. 68: no. 5, p 935-47.

Yapp, Creighton, 2005-07, Prof. Dept. of Geological Sciences, Southern Methodist University, Personal communication.

Zachos, James, et al., 2001, "Trends, Rhythms and Aberrations in Global Climate 65 *mya* to Present", *Science*, v. 292: P 686-93.

Zahn, Rainer, 2002, "Milankovitch and Climate: The Orbital Code of Climate Change", *JOIDES Journal*, v 28: p 17-22.

INDEX

A

acidification 88

aerosols 13

albedo 14, 15, 23, 24, 40, 43,
45, 48, 54, 55, 64, 74,
102-104, 112, 116

algae 7, 30, 57, 80

Anazazi Indians 93

Antarctic 36, 42, 52, 68, 69, 91,
93, 110

Archean. 4, 79

archeology 66, 78, 94

Arctic Circle 98

Atacama Desert 39

atmosphere xi, xiv, xvii, 9, 12,
14, 25, 46, 62, 64-66,
68, 72, 75, 80, 86, 89-92

atmospheric circulation 57-59,
116

axis of rotation 22

C

Cambrian xi, 4, 82

Carboniferous 4, 72, 85

carbon dioxide xi, xii, xix, xxii,
2, 4, 9, 13, 14, 21, 25,
27, 30-38, 44, 45, 47-52,
56-58, 63, 64, 68, 69,
72, 75, 76, 79-89, 91,
92, 94, 97-99, 102-104,
106, 109, 111, 112, 115-
117, 119-127

chemical weathering 14, 47-49,
58, 91, 92, 102, 103, 116

clathrates 34

climate drivers xxi, 14, 59, 63,
69, 76, 99, 109, 113,
114, 116, 117, 120

Climate Indicators 9, 65, 78

coal 30, 38, 74, 111, 123

continental collisions 44, 47, 48,
89

continental drift xi, 45, 70, 73,
88

continental plates 43, 44

continent locations 14, 43, 47,
81, 82, 84, 85, 88, 89,
92, 116

convection cells 43

corals 9, 66, 67, 77, 78, 82, 87,
89, 90

cosmic rays 10, 14, 47, 59-64,
66, 98, 104, 106, 116